Eddie LeRoy

INSIDE THE TV BUSINESS

INSIDE THE TV BUSINESS

Paul Klein

Alan Landsburg

Norman Horowitz

Rowland Perkins

Dennis Swanson

Lawrence Fraiberg

Margaret Loesch

Joe Barbera

Don Ohlmeyer

Richard C. Wald

Sonny Fox

Edited by Steve Morgenstern

Sterling Publishing Co., Inc. · New York

Contents

Foreword

I broke into the broadcast business many years ago as a page boy at NBC. At the same time I was an undergraduate at Fordham University with a major in communications arts. This seemed to be a good combination: theory in the halls of scholarship, and practical knowledge in the marketplace. And to some degree it was true. But the learning areas available to a page were limited, and the academic theory proved to be somewhat detached.

However, one day the instructor invited two guest speakers to address the class. One was a network executive, the other an actor. I don't recall much of what the executive had to say (he had apparently mastered the principles of neutral-speak early in the game), but the actor made a lasting impression. Instead of lecturing on the basics of Stanislavskian method, he told us how difficult it was to get an acting job—the endless making of the rounds; the need to discover who really is the person to see at agencies and networks; the techniques for endearing one's self once a job is secured. It may not have been a stroll through the groves of Academe, but it certainly was a useful guide to survival in the real world. At that moment I became a fan of the informal, straight-from-the-shoulder form of communications by seasoned professionals.

Over the years the business has become increasingly demanding. The stakes are high and the competition relentless. And to add to the fun, the nation's viewers grade our performance on a daily basis. To achieve success it is essential to know how the process really works. That information can only come from those who make the decisions and set the ground rules.

I think Sonny Fox has assembled an extraordinary group of executives, insiders if you will, who talk openly and candidly about their roles in the medium. Their comments and insights will have value to everyone interested in this fascinating industry.

EDWIN T. VANE
Vice President, Network Program Affairs
ABC Television Network

Introduction

Television programs are highly perishable.

So are television programmers.

Since the lectures which form the basis of the chapters in this book were presented, at USC's Annenberg School in the spring of 1978, three of the ten lecturers have changed positions. Paul Klein has left NBC to go into independent production. Larry Fraiberg has left Metromedia. Dick Wald, previously at NBC News, and at the time of the lecture, Executive Assistant to the President of the Times-Mirror Corporation, has now become Executive Vice President, News, at ABC.

Although 30 per cent of the executives who lectured have been displaced within the year, the realities that lie behind the business of television remain constant. The decision-making processes and the economics that Paul Klein describes in his chapter have to be dealt with at each of the networks. Paul's leaving does not change the realities with which Fred Silverman has to deal.

The reason that each network's program schedule bears a close resemblance to the others, and the reason so many network executives move from one network to the other (Fred Silverman has completed the circuit) is that they all must deal with the same realities.

This book is about those realities. If you are thinking of getting into the business, it is well you know about them. If you are in one part of the business, it is helpful to know about the other parts. If you are one of the millions of consumers of this most affecting of all media, it is important that you have a better understanding of the underlying realities with which everyone in the business must deal.

My thanks to Dean Fred Williams and his staff at the Annenberg School for their cooperation in presenting this series of lectures, and of course, to the remarkable group of professionals who shared with the utmost candor the lessons they have learned over the long and sometimes bloody years.

SONNY FOX

Sonny Fox, moderator and organizer of the lecture series on which this book is based, has enjoyed a multifaceted career spanning over 30 years as a television performer and producer. He hosted *$64,000 Challenge* and the long-running children's show, *Wonderama*, among others. He was associate producer of the late-night *Tomorrow Show* on NBC-TV in 1973-1974, and served as Vice President, Children's Programming at NBC in 1976-77. Fox served as Chairman of the Board of the National Academy of Television Arts and Sciences 1970-1972, and is now an honorary trustee of the Academy. Today he is active as an independent producer of television programs.

Programming

Paul Klein

An outspoken and incisive observer of the broadcasting scene, Paul Klein served as Executive Vice President, Programs for NBC-TV from November, 1977 to March, 1979, during which time he made the remarks which follow. He left NBC in March under an agreement to produce programs for NBC Entertainment.

Klein holds a B.A. degree in mathematics and philosophy from Brooklyn College. He began his career in advertising in 1953, and served as research manager for the Doyle Dane Bernbach advertising agency in New York 1955-1960. He joined NBC in 1961 as Supervisor, Ratings, and rose to the position of Vice President, Audience Measurement by October, 1965.

In August, 1970, he left NBC to found Computer Television Inc., which became the first independent pay-per-view television company in the world. After establishing systems in most of the major markets in the United States and Canada, the company's largest investor, Time Inc., bought Klein's interest in CTI and its sister organization, Home Box Office. He then returned to NBC to work in the programming area.

In addition to founding CTI in his time away from NBC, Mr. Klein was a contributing editor to the then fledgling *New York* Magazine. He has also written articles on television for *TV Guide* and scholarly media journals, and served as a consultant to PBS, CPB and the Ford Foundation in the area of television programming and scheduling.

NBC and the other networks follow a particular schedule every year in selecting the programs which you will eventually see on the air. One of the reasons I'm in Los Angeles in March is that this is pilot season. Actually it's a little late this year, but pilot season is when you make sample programs of the series that you plan to put on next year. We make about 45 pilots a year. Most of them are done in the first quarter of the year, and then we see them around April and May, when they've finished shooting. It's going to be a little late this year, because of the rain, but usually we make our schedule by the end of April for the following season.

In the old days we used to make our pilots in the winter, on the West Coast, mostly in studios, and schedule around March, or a little earlier. Washington's Birthday used to be the classic time to come out with a schedule. Then the advertisers would start to buy the programs, which would remain on the air 39 weeks in some cases, followed by 13 weeks of either replacement or repeat. Then it got down to 26 weeks and then 22 weeks. Now 22 weeks is considered a full year, with 22 repeats and 8 weeks left over, in which we all "stunt"—put in specials, things like that.

Today the advertiser is not waiting to buy anything. He doesn't own anything, he doesn't possess anything, he doesn't promote anything. He buys in what used to be called the "magazine concept." He buys spots in programs, so it doesn't make any difference that we're coming out so

late. Procter and Gamble will buy something here, something there, not a salacious program here, not a salacious
program there. Everybody hunts for their little target audiences and buys a little bit of this, a little bit of that.

Now we come out with our schedules around the end of
April. It will probably be the beginning of May, because
everything's coming in late, and we're trying out series on
the air.

I just came from seeing a pilot which we're actually going
to put on the air. Right now we are making four episodes, of
which this is the first. We will stick it on the air and see
what happens with it. We're shooting it as a three-camera
tape show, which I'll explain later, in our own studios in
Burbank. This is a Jimmy Komack production called *Roller
Girls*. It's an idea of ours which we gave to Komack to produce, and it's about a team of girls who are in the roller
derby. They're the expansion team, in Pittsburgh, and
they're called the "Pittsburgh Pits"—that's the entire funny
line of the show. They all come from different teams, and
one is just a rookie. She's the nice-looking one. She was a
carhop girl on roller skates and the manager of this team
saw her and booked her for stardom.

There are five girls living together, struggling, and doing
the roller derby. The first episode will appear at the end of
April, the second episode beginning of May, and so forth.
We'll probably play it against weak competition to get it
sampled. Then we'll make a decision about the show on
what we see, and what the audience reaction is. I looked at
it, and right now I'm ready to kill myself, but that was the
dress rehearsal, and we're still revising scripts.

Right now, in March, we're shooting a lot of pilots simul-
taneously. There are a few ways to shoot a pilot. We make
comedies in three-camera tape, three-camera film, or
single-camera film, and we're doing about 11 comedies
right now, maybe 12. We do what's called "long form" for the
dramatic shows, meaning we make a two-hour movie as a
pilot for an hour show. The shows we're making right now
include one called *Capra*, about a lawyer-detective. Natu-
rally, you have to have one detective show. We have two girls
from the Old West, *Lacy and the Mississippi Queen*. We

have a lot of other properties, some of which I'll name as I go along.

Casting

We're casting now too. We work very hard casting. We have never done this before, because it wasn't as competitive a business before, but we struggle now to get the right people. I think ABC last year was guilty of a little restraint of trade. They actually signed the people before they had the parts, so then they controlled the people for the pilot season. Only ABC could touch these people, so they had the pool, and then they placed them where they wanted them. This year I don't think they will do that.

Theory of Least Objectionable Programming

All right, so let's say that we have all our pilots shot, and we have the shows that we can bring back from last year. Now we have a schedule in front of us, and we want to fill it with programs that will succeed, in television terms. Programs that will get a competitive audience.

What does that mean? I have a theory, which you may have heard before. This theory says that the biggest star in America is not any actor, or show. It's television. People watch television, they don't watch programs. And, they are there every day. The number of sets in use, the number of homes using television on a particular day, is identical to the year before on that day, and the year before that, and the year before that, and all the way back to 1960. In 1959, there was a big dip, there was the quiz scandal. And prior to that they were the same. And they're always the same. When you're available, you watch television. When you're not available, by definition you don't. People say things about television, "It stinks," "It's no good," and they say these things because they watch so much that they watch things that would not normally be at their level of taste.

The theory evolves this way: A man comes home at night, and he's tired from a day's work. He sits down, he eats his food, and then he sits back and says, "I think I'll watch a little television tonight." He says this in a voice as if he had never watched television in his entire life, and this is the night he's going to do it.

Then he goes to the television room, sits down, and turns on the television set. It's on Channel 3. There is nothing on Channel 3 but snow, but that is where he left it the night before at 1:30 A.M., when he was searching around for something to watch instead of going to bed with his wife. He then looks at the snow for a while, "Ah, it's not bad, it was better yesterday," and then he turns around clockwise, and he goes to 4, he watches that for a while, he says "Oh, what s – – –," goes to 5, watches that for a while, says "Oh, what s – – –," and then he turns to six, more snow. He watches that a little longer. "It's better than 3 snow, but it ain't so good," and then he goes to 7. And then he watches 7 for a while, then he watches 8, more snow, lingers there a little bit, 9, and he goes around faster faster faster faster and he settles on a program, which I call the least objectionable program. And he watches that. Then he makes the least objectionable choices all evening, and that's how television is viewed. At the end of the day, he goes to sleep, wakes up the next morning and tells his friends that he never watches television, and that it all stinks. Those two sentences can't exist together, "he never watches television" and "it all stinks," but the reason is that he's watched all this television that does not live up to his preconceived notions. It can't, because it's a limited spectrum medium.

We exist with a known television audience, and all a show has to be is least objectionable among a segment of the audience. When you put on a show, then, you immediately start with your fair share. You get your 32-share, and I'll explain shares later, but that's about 1/3 of the network audience, and the other networks get their 32 shares. We all start equally. Then we can add to that by our competitors' failure—they become objectionable so people turn to us if we're less objectionable. Or, we could lose audience by inserting little "tricks" that cause the loss of audience. NBC has been very good at doing these little tricks lately to cause tune-out. Thought, that's tune-out, education, tune-out.

Melodrama's good, you know, a little tear here and there, a little morality tale, that's good. Positive. That's least objectionable. It's my job then to keep my 32, not to cause any tune-out *a priori* in terms of ads or concepts, to make sure that there's no tune-out in the shows *vis-a-vis* the competition, and to look and see where I can take advantage of their vulnerability.

Now, when we have a schedule on the air, some of the programs are doing well, and some of them are not doing well by these criteria. We then throw off the ones that are not doing well, and we put in new ones which we think will do better, that will not only get a bigger number (their fair share of the audience), but maybe even try to be a hit, and provide a lead-in, because, all things being equal, people prefer not to get up and change the channel after each program. So if you have a good lead-in, a show that people actively want to watch, they're likely to passively watch the following show on the same channel. In fact, in a big part of the country you still have to use the fine tuning when you switch channels, so it's always good to have a comfortable and comparable program following. Doing this is an essential part of my job.

Segmented Audience Schedule

There are, of course, other criteria, other strategies which come into play in deciding which programs to put on the air. At this point, NBC has a schedule (and so has CBS) that has aged. ABC is killing us with what's called a segmented audience schedule. They've done what rock radio stations did back in the late 50's and the early 60's, where they programmed the same kind of sound all day long, and that sound is targeted towards a particular segment of the audience, specifically the young people. ABC has taken this concept and brought it to television programming by not trying to appeal to everybody, but rather concentrating on one segment, the young people. This gives them high ratings, since the control of the set is largely in the hands of this young audience.

When I was at NBC originally, back in the 60's, we out-programmed CBS. We left their programs with just kids and old ladies. Those were very high-rated shows, because kids and old ladies live in different homes, and the Nielsen measure is a home measure. But the advertisers did not buy the kids and old ladies because they don't consume much product. So CBS eventually threw off *Beverly Hillbillies, Petticoat Junction, Red Skelton*, etc., etc., etc., their highest-rated shows.

What we're attempting to do against ABC this time, and it's a very hard job, not done overnight, is to skim off the top of the audience scale. That's what we've been working on, comedies that do that. *The Three Wives of David Wheeler*, etc., etc., etc. Don't get the kids, be a little more sophisticated than that and skim off the top. That's the target that we're aiming for. Now, we're not going to succeed all the time. We're probably not going to succeed even one third of the time. But if we do 20 or 30 per cent of the time, we will have one fantastic year, and we will do a job on ABC, enough to lower their rating points, and pick up the most saleable part of their audience. That's what I'm looking to do, targeting the programs to do that, and it's not easy.

Ratings

What I'm talking about here is not only improving the rating points that a show gets, but selling the idea, from the advertiser's point of view, of targeting his television dollar towards shows which might not get the biggest numbers according to Nielsen, but draw the kind of people who are likely to buy this man's product. This whole business about ratings has people confused. You see them in the papers, but you probably don't know what they mean, and very few people have any idea what the difference is between a rating as opposed to a share. Let me try to explain it briefly.

To start with, a rating has nothing to do with judging quality. A rating is a measure of circulation—the circulation of the commercial position in the program. If a show gets a 20 rating that means that 20 per cent of the total number of homes with television sets in the United States flowed through the average minute of that program. And there are

73 million homes with TV sets in the United States, 20 per cent of those is 14,600,000 homes. That's the circulation of the program, or the circulation of the commercial in the program, which is the only thing that matters. The share is more important, because that tells you what percentage of the sets actually in use at a given time were tuned to your program, so it's how well you are doing competitively. If I have 14 million homes, a 20 rating, that may not be as good as it should be, because my competition each may have 25 ratings. So the share tells me, of all the homes using television, how many tuned to my program. Did I get a 25 share? 30 share? 32, I said, is my rightful share going in. If I got lower than the 32, then I'm not as competitive, meaning my competitors are getting bigger numbers. If I get higher than the 32, I'm depriving them of audience, and I'm selling extra circulation on my program. Share is the most important thing. Rating is what you read in the papers. Rating appeals to the newspapers, because it sounds like it means quality. The rating does not indicate whether a show is terrific or lousy. If the rating is high, then a lot of people watched it, and it might have been good or bad. If the rating is low, the people who didn't watch it wouldn't know that it was no good. The people who watched it might not have liked it, but that's all right, they were in the circulation. It's only circulation that we're measuring, nothing about liking or disliking. And we try to make these ratings very accurate because we live and die off of them.

But you're living and dying off them because the ad agencies and the clients will spend or not spend their money based on them. So when, say, an advertiser gives you a 31 share, it is an indication of what they are willing to spend for the commercial. They believe that that's the share of the audience that will see their commercial. Now, if your average share over a season is substantially below that of your competitors, you have to project a believable share in the future upon which you will base your prices. You've got to be believable, otherwise they won't buy at that rate. And if you come in third, with a fairly substantially depressed share, then you're going to get less money for time on your network in the future than your competitor's going to get.

Hyping the Sweep

Ratings are taken for the networks all year long, all over the country. In addition, three times a year, in February, May and November, there are "sweep" periods. Everybody's confused about the sweep. A sweep is a rating of the local station. There are two kinds of ratings. There's a national rating, measured every day, every minute of the day, by Nielsen. Nielsen measures electronically, that is, it sends an impulse to a pre-selected, randomly picked sample. The impulse is a telephone call to a telephone attached to the television set. That telephone on the set is doing the following: it is saying, I am on or off, and if on, on what frequency. That frequency is matched with the program that is on that frequency in that market, and Nielsen then records it as a home watching that program. This national sample is measured constantly. A home in the sample gets measured every 8 seconds.

Then there are sweep periods, where the local stations get measured. The local stations don't get measured with this very sophisticated electronic measure I described, except in New York, Chicago, Los Angeles and San Francisco so far. They get measured by a sample of homes which are sent a diary for a week. Nielsen sends diaries to a sample home in each county in the U.S., and then they add up all those diaries and come out with local ratings in the station's area of influence (coverage area). There's another rating service that does that too, called ARB.

The networks have affiliated stations that are dependent on them, so it was important to what I used to call "hype the sweep." That is, to put on our best movies for the four weeks that were measured locally. So we would put *The Godfather, Gone with the Wind*, whatever, in the particular sweep period, and as a result we'd get an abnormally high rating. Networks don't really sell on the basis of the local ratings, but the local stations sell their local ratings. NBC, with a good movie package, was always able to hype the sweep better than its competitors, but now the others are hyping just as well. CBS just had 32 days of February promotion, that had nothing to do other than to hype the February sweeps. They threw in everything they had, every penny, just to satisfy their affiliates, which are being raided by ABC. Then ABC did the same thing. You'll see the same kinds of programs in every sweep. A challenge of the network stars, a battle of the network stars, a challenge of the beauties, whatever, whatever hype you can do. Very expensive movies and things like that, and a lot of miniseries.

May is the second sweep period. May used to be a month where everybody played repeats. The ad agencies and the independent stations wanted to get measured then, so the networks would look weaker and the independent stations would look better. When I came back to work at NBC I started hyping the May sweep. We put in original programs, everything and anything we could find, and we did very very well, and that sort of saved NBC for its 50th anniversary party. And then the other guys started to hype them the next May, and now this is the third May since I'm back, and we're all hyping like crazy. New *Happy Days* are being made for May. We're putting in *Wheels*, the Arthur Hailey novel done as a miniseries, ten hours of it. CBS is putting in *The Dain Curse* by Dashiel Hammett as a three-parter. Everybody's putting in big movies. We've got a movie with Louise Lasser we're putting in. We have original programs, a new Hanna-Barbera hour, new shows like *Roller Girls* and *Joe and Valery*, we're playing those in the sweep. We're just loading the sweep, and the other guys are hyping as well. And May has now become the show business month of the year. You have to stay home.

Fake Realism

One of the qualities I look for in a program to be a commercial, popular success is what I call fake realism, phony realism, the illusion that you're getting something meaningful. You then overlay that with what I call trash. That's the ideal program to get a woman between 18 and 49 to watch, and that is the ideal market. That's what the book publisher looks for all the time. It's not Camus, but it is the ideal content to fit the nature of a "time waste," which is what people use media for.

So we're looking for that kind of thing. People then read

into it that it was really a work of art, or whatever. It might even be a work of art, who am I to judge a work of art? *Casablanca* is now considered a work of art, as a movie. There's nothing in there of any consequence, but it's considered a work of art. When people study it, they tear it apart. It was just an exploitation film with two stars. We are looking for that kind of thing, and in the future someone will say that *79 Park Avenue* was a work of art, or they say *Rich Man, Poor Man* was, or *Roots*, or something like that. Fake realism like *Roots* I think is one of the ideals.

Miniseries

The miniseries is a recent development in the packaging of this kind of fake realism. When I came back to work at NBC, which was just two years ago, *Rich Man, Poor Man* was just going on. We had done a few similar things prior, in the late 60's. NBC needed a new, fast audience. We were dying, all our programs were dying, and I felt we could get an immediate audience from miniseries at that point. People were interested in the form, irrespective of the content. You could also develop a program quickly, because there was an existing piece of work, a book, and the highest quality professional people wanted to work on it. The best writers, the best performers. A performer would prefer to work on a miniseries rather than on a regular series because the mini had a finite period of work, and the regular series required a commitment, possibly for four years.

But producing miniseries was expensive. It wasn't *that* expensive, but the prices kept escalating as competition set in.

But the reasoning behind going with the miniseries was to get right back in the ballgame, and we did get right back in the ballgame. We were hoping that we would then develop series that would carry this momentum. Then we could be more selective in our miniseries, if we had regular series going. But we didn't develop hit series as we had hoped, so now all we have is the miniseries and we're still struggling to get regular series that lead in and provide extra audience to these quality miniseries.

There's been a lot of discussion on the economics of doing miniseries, and a lot of misunderstanding. Somebody got the idea that the miniseries cost so much to make that even though they get big ratings they don't produce a profit.

We buy a miniseries and we pay a certain amount of money for it. The cost ratio runs about $550,000 an hour. It's a generally accepted number, about $1.1 million for a two-hour segment. That's a fairly high figure, some of them were a little over that and some of them were under that, there were some ranges, and now the ranges have gotten even bigger. Presently it's about $600,000 an hour. And then we sell time in it. Had I put a regular program in there, a program we developed and bought 13 or 22 episodes of, I would have paid $400,000 and needed two runs. My chance of getting a 30 share, a little under my fair share, would be one in eight. So that if I put a regular program in there, try to develop something, I would automatically take a huge loss, most of the time. And if I had a lot of regular development programs in the schedule, I would take a lot of losses. So I reasoned, as a mathematician, that if I put in a miniseries, the ad agencies would immediately give me at least a 30 share. (Actually they gave me a 31.) Not only that, but they would give me a quality audience that they would pay a little extra for.

Another advantage is that instead of playing an hour, or a 30-minute program, I'd be playing a two-hour program, and I'd get more commercial minutes in it. I get 7 minutes per hour instead of 6 per hour in the regular program, because there's so much other non-program material in an hour that I don't have to include with a continuous two-hour program, so I can play a few more commercial breaks in there.

A unit, I should explain, is a 30-second announcement. When I first started in this business it was God's will that there shouldn't be 30-second announcements, I remember, because a minute was the right form that God had sent down to earth from heaven.

So, if you have 7 minutes an hour, you have 14 units an hour. If you have a two-hour program you've got 28 units, and I can sell a 31 share of that quality for $50,000. I can sell 28 units at $50,000, so I gross $1.4 million. Out of that I pay for the program, which I just told you was 1.1 million, and I still have a repeat waiting for me.

And I pay the stations out of that, station comp. I buy their audience from them—that's how television works. We pay the stations a commission, which is called station compensation, or station comp. I go to them and I say, if you carry my program, and give me your audience for that program, I will then sell that audience to some third party and give you a commission. That's how networking works.

Then we also pay a commission to an advertising agency who places the time. So I probably have a break-even on that first run, and I have a repeat, that I now have home free. I have a two-hour show that I can play some summertime, and gross maybe $30,000 a unit, and $30,000 times 28 is $700,000, no program costs now, just the commissions, and I'll make about half a million dollars. So on this venture I can make about half a million dollars. In the other venture, the regular programming that I was thinking about running, I would lose lots of money. And besides, the miniseries would look like it was quality—and just like that it's the golden age of television. Besides the economics of it, it's a better bet, it's better for the public, it's a good thing to do.

So that was the reasoning. But the first time I played a repeat of a miniseries it went down the tubes, and then CBS had *Helter Skelter,* which was a high-rated show, and that went down the tubes in repeat. They got very low ratings. The repeat was not competitive. We've only tried repeating in March and April. We just took the strongest miniseries we had, *Sybil,* and played it this week. It got a 26 or a 25 share but we made a lot of money on it. Of course, that was a low-cost miniseries to start with. But we lost competitively, and that cost money in the future. When we get a 25 share, we lose our rightful 7 share points. They get distributed to the other two guys, and they then convert that into dollars *in the future.* So that's very important.

It says in the paper that we have preempted regular programming to put on miniseries, but we really haven't. The problem they're talking about is that when you buy a certain number of episodes of a series, you pay for that number, whether or not you run them all. So if you put in a special, or a miniseries, or whatever, instead of that regular series, you're paying double costs for programming that particular period of time. You pay for the original show and the new special show, and it's tough to make a profit when you've got to absorb that kind of costs.

We didn't do that. When we made a big thrust toward miniseries, we took all the programs that were working on a particular night and we moved them to another night, and we left blanks there so we could play our miniseries. So we're not preempting, we have not done any preemptive miniseries. We don't have a regular program for certain time slots, our regular program is a miniseries, just like the BBC would have.

Program Costs

Program costs vary depending on the kind of production required for a particular type of show, and what we expect to get out of it in terms of developing a continuing property, getting big one-shot ratings, and so forth.

This year at the networks we're paying up to $50,000 for a development script for the pilot of an hour-long nighttime series. We go up that high if we want a particular writer, or even a writer-owner on a project like this. Otherwise we pay in the range of $35,000. We've had pilot scripts for $20,000 from a new person, somebody trying to work his way up.

The license fee is what we pay for the right to run a program, two times usually. There are three types of half-hour shows, beginning with the three-camera tape, which is the cheapest. The fee for that is up to about $180,000, give or take some, and shows do come in cheaper sometimes. That figure can go up to $200,000, depending on the casting and things like that.

Then there's the three-camera film, which runs about $210,000 to $220,000. Film costs more money than tape, and the editing is harder, so it becomes more expensive. Then there's one-camera film which is the most expensive. Here you start using exteriors, and it might run as high as $240,000 or $250,000. At $250,000 it starts to get uneconomical, but a show with a big cast, like *Soap,* would be $250,000 or even more per half hour. Plus *Soap* is not a repeatable show, because it's a continuing show, so you'd have to repeat it in sequence, and those things die that way. So a show like that starts to become a losing proposition. $220,000 is still worth it, and it's not hard to see how the numbers are assembled.

How much we pay for the show is dependent on our rightful share of audience. If you figured it based on getting 14 million homes as our rightful share, and you sell those homes for $3 a thousand, we would sell it for say $40,000 per unit, and we'd get 6 units in a half-hour show, so we'd gross $240,000. If we had a hit we'd gross a lot more. Therefore we'd break even, and we'd get a free repeat. We would make all our money on the repeats. The way we allocate the cost is we take 70 per cent for the original, 30 per cent for the repeat.

On a show that costs us $200,000, we'd allocate $140,000 as the first run cost, we would gross $240,000. We'd make $100,000 or $80,000 for a show that is performing adequately in its first run. For a show that dies, we're not going to get as much money, and then we have to replace it because when it dies, it does two things. It hurts us, and it helps the other guy. It makes hits out of the competition. The most important characteristic for any show is weak competition. News is the best kind of competition. If you have a documentary opposite you, you know you're going to have a hit, because while people love documentaries, and when you interview them, they want to see more of them, they didn't say they would watch them. They said they would like to have them available just in case they're in the mood, and they're rarely in the mood.

For an hour series now we do 16-millimeter shows, and they could be as low as $250,000 or $275,000. We've bought shows for that money, and as high as $450,000. A *Bionic Woman* might run that high, with the special effects and all that. So the mid-range is about $360,000, something like that.

For long forms, meaning a two-hour movie that you get to play twice, an independent guy like Alan Landsburg would get something a little over a million dollars. We'd also pay the breakage, which is the extra cost from star casting if he can get some big name and has to pay $100,000 to get him. We'd kick in the overage, over a certain amount of money, so we might run $1.1 million, $1.2 million, maybe $1.3 million.

If it's a miniseries you take the same $1.2 million and multiply it times three if you have it three nights. Those are all comparable figures, though, within the ranges I'm quoting, because they all come in at about the same dollars per commercial unit. The numbers are established on the basis of what you're going to get from the marketplace for a return on your investment.

And yet we'll pay $8 million for a movie, and not even a blockbuster movie, a theatrical movie that's been pre-sold to the public. I mean, ABC paid $14.5 million for *The Sting*. Of course, for nearly $15 million they get five runs. You buy as many runs as possible, just for accounting purposes, because you then delay the write-off of the cost of the movie. If you allocate 40 per cent against the first run, it's $6 million against the first run, and you break even on the first run. Then you put 20 per cent against the second run, and 10 per cent against the third, fourth and fifth runs. You may not even take the fifth run, and you write down that loss, but that's five years hence. By then other people are working there, let them suffer.

The exception is something like *Gone with the Wind*, which we bought for one run, for a blockbuster. We bought it for the sweep. And even then we made our money on it. We paid something like $6, $7, $8 million, I forget what it was, and we got it back because we charged a premium price. The advertisers gave us the 65 share, which is double what we would normally get, so we double all the dollars. And the local stations, the stations we own (O and O's) besides our affiliates, charged extra money for the spots around it. They got extra money for the news that followed it, all these things were extra. In addition, it was in the November sweeps, so it stayed in the average from the November sweep to the February sweep. We got paid for it for three months. That's why you buy known blockbusters.

ABC now has a lot of discretionary dollars around, because they've had high-rated series, so that they don't have to protect themselves against attrition of their own series. The guy with the highest-rated series becomes the most vulnerable network for the following reason. As his series starts to fall from a 40 to a 35, he cannot replace it, he's still got a high-rated series, and the next year he's down to a 32, he's now in the marginal area where he can't replace it, but he knows in his head he's got a 28 share season coming up. Still he can't take a chance and put on a new show, because he's only going to get a 26 on the average, so he takes the chance with the 32, and becomes the most vulnerable network. NBC had that once and CBS had that once. At some point three or four years down the road, they're going to be in a tremendously vulnerable position. CBS had that and Fred Silverman switched to ABC, at just that moment, when everything fell apart, and now you know where he's at!

It's really a tense problem when you have that—high ratings and known vulnerability. I think what ABC is doing now is a very good thing, which the other networks did not do, because they're bigger companies. NBC would throw off the money to RCA and pay the stockholders. But ABC instead has bought *Jaws, The Sting, Saturday Night Fever,* known high-rated shows. They've taken their discretionary money, something over $100 million, and have bought known winners. And they will then protect themselves for many sweeps to come. You'll see *The Sting* in November, and you'll see *Jaws,* and not only *Jaws* but *Jaws II,* the sequel. But then again we bought things too, you know. *Airport 77,* and *Airport 79,* and *Super Airport,* we keep buying into the future.

Loss of Viewers

QUESTION: I was reading in *Time* magazine and also in other recent publications that there's a real concern for the loss of viewership that's been going on. They attributed it to two things. One was that they felt some of the programming was really lousy, and then, in a more scientific way, they thought that it might be the fact that housewives were working more. I wondered what your theory might be in regards to that.

PAUL KLEIN: First of all, there is no loss in viewing. The programs were lousy last year, by their criteria, and they measured the loss against last year's audience. Programs are always lousy to *Time* magazine. You pick up any *Time* at random, and they'll say the programs are lousy. Magazines are a competitive advertising medium, and that's all they can say. They're not going to say it's good, it's the greatest thing, put all your advertising money into television. No, they say it's lousy, don't put your money there. Television, being magnanimous, never says that *Time* magazine's a lousy magazine, or has low circulation compared to say, *TV Guide*, that they stink compared to *TV Guide*.

Time magazine's always equating our circulation, the number of sets in use, to how good we are. It has nothing to do with it. Circulation has nothing to do with content, it has to do with what the medium is. The same lousy programs you read about in that article were also on in January when the sets in use went up, and in February it went up again. The same lousy programs are now causing the television usage to go up. It's just an adjustment in the Nielsen sample, it has nothing to do with anything. The differences were one point here, two points there.

The *Wall Street Journal*, which I love, has really good theories. Their theory was that the television viewer was dropping out because there are too many stunts, too many programs not on in their regular time period. And that was presented as a theory by a well-known person. But he never thought the thing through. In daytime, where viewing was off the most, that's where the programs are on with terrible regularity. They're on every day, the heavy, heavy troubles are on every day, without fail, and the sets in use are lower. Then came the theory on top of that that women are working, which is really true.

My theory of lower sets in use, the *minor* drop in sets in

use, is as follows. What has really happened in the population is that for years there has been a slow change taking place in the pattern of family life. People have traditionally gotten married at a certain age, around 20 years for a woman—and remember women are the basic audience for television. And now these women are not getting married at 20, they're getting married a little older, and then, even when they get married, they don't have babies as they used to. In the past when a woman had a baby she went immediately into the television market, or if not immediately, in the seventh month when she left her job. She came home, had nothing to do, avoided the television set, avoided, avoided, avoided, finally when she was 7 1/2 months pregnant she sat down and stayed there, maybe through three babies, till the third one got to be 19 and moved out.

Now, there's a longer period before they have babies, and many women aren't having babies at all. Or they're working right through, going right back to work and putting the baby someplace else, or carrying them on their backs to the office. They're in the labor force, not in the viewing audience during the day. There has been that kind of attrition, particularly in the daytime, although not in the *Today Show*. *Today Show* was up, as a matter of fact, and so was *Good Morning America*, because there were more young people in the audience. Then the daytime numbers fell off and they didn't come back again till about 10 o'clock at night, and the *Tonight Show* time period is high now. *Tonight Show*, *Saturday Night Live*, have record high figures of sets in use during those late nighttime periods. So those people moved into the later time periods. But in general the total numbers of usage per home fell about 2 minutes a day. The average home, instead of using television for 6 hours and 47 minutes, now uses 6 hours and 45 minutes, but there are more homes using it that 6 hours and 45 minutes.

The Star's Percentage

QUESTION: There's recently been a rash of stars and writers who own part of the profit in shows, and they complain they've been on the air for eight years, and there are never any profits. Who's making the money?

PAUL KLEIN: The stars are making the money, but they want to get huge salaries, and then also want their share of the profit. And there is no profit for years. Let's take a show for demonstration purposes, *Rockford Files*, where you have James Garner, his agent-producer, Meta Rosenberg, Steve Cannell, all as participants. Steve Cannell's the writer-producer, the hyphenate on the show. Those three people are owners. Universal pays Jim Garner a lot of money per episode, pays Meta Rosenberg a lot of money per episode, and pays Steve Cannell a lot of money. It costs them, say, $450,000 to produce the show every week. NBC pays them on the average say $400,000. They "lose" then, $50,000 every episode. The next year they get paid more but their overhead sometimes may be more. They build up the negatives. They wait for the show to be over, then they sell those negatives into syndication or late night. This is how they look to recover their money. This is a steady process, the money coming in from syndication deals on older shows covering their deficits from current production and providing them a profit as well.

So now *Rockford* doesn't go into syndication until maybe the seventh year. I think they're allowed to syndicate now, even though the show is still on the networks. So that the stars haven't seen their percentage of the profits yet, because there are no profits yet. When the program is put into syndication, they get $200,000 an episode by selling it to the individual stations in the United States. So that's $200,000 an episode and they've got four or six years worth of episodes. And they make $150,000 there; that's the profit that the producer's talking about. And that's what he wants, because they've just recovered the $50,000 they lost on the first run. But at that point, there's a distribution fee for syndication, which can run to 50 per cent, and this is where all the talk comes from, of actors like Michael Caine suing people. It's that distribution fee, and how it's determined. It was determined *a priori* at the time Jim Garner made his deal with Universal. Now a guy is not going to worry about that fee at that point, he's going to worry about his up-front money. So he doesn't worry about the distribution fee, until it comes time for Universal to start getting theirs. He then says, "You cheated me." You know, he might be the one that cheated them initially by taking a lot of money up front, but they financed this thing and they're waiting for their what is called "back end." For that money down the road, from that steady syndication process. And therein lies the pulling and the tugging. The more successful the show, the more the tug. The star's already made a lot of money but he's in for a piece of the profits, which might be an extra $15,000 or so.

If I'm Jim Garner and I'm willing to take a risk and wait for that back end, I will work for $10,000 a week, and take a bigger piece at the end. But he'd never do that, he'll take his up-front money. In fact, after the second year they may go and sell their back end portion to a bank so that they have more money to build that big house on the top of the hill.

SONNY FOX: There is, however, one more thing you really have to know. If you're working as a producer on the Universal lot, and NBC pays you a million dollars to do a two-hour movie, Universal will take off the top, for overhead and other aspects, anywhere from $200,000 to $300,000. And that's for the use of the offices, the studios, the makeup, the facilities, and all the other things. So it isn't that Universal doesn't get any money until they get recoupment through syndication. That $50,000 first run deficit includes $200,000 or $250,000 which Universal has taken.

Violence on Television

QUESTION: What about the general trend towards violent programs? For instance, the recent incident after the showing of *Born Innocent*?

PAUL KLEIN: You're talking about a made-for-TV movie,

Born Innocent, which had a scene about a girl being molested by other girls in a women's detention home. And that supposedly caused one group of people in San Francisco to perform this same thing on somebody else. And they got it from television, according to some people. About 16 million homes saw it, but television was so weak that it only caused that one incident. This case went to court, where somebody was trying to extract money from NBC because of the causal relationship. NBC won, but before the case was decided, it was all over every newspaper that there was a causal relationship.

Now there's no question that we're dealing with a very powerful medium. The medium itself is powerful. The content is nothing, but you're dealing with power, and if you say a causal relationship has happened, and then you go out and sue, it'll get in the newspapers. The newspapers are a competitive medium, and they sell newspapers by writing about television, because they know everybody watches television and thinks something is happening to them. I always thought the best article would be, with a big banner headline, "Television Is Good for Your Eyes." It would sell out immediately because you'd want to know how good it is for you, because this is something you use. It's like people who smoke marijuana want to read articles about how it is good for your bloodstream or for your sexual drive.

The case went to court and they decided there wasn't a causal relationship. They decided for NBC.

I remember lecturing the day after ABC ran a movie with Raquel Welch and Burt Reynolds in which a bunch of kids go to a dock and burn somebody, and in Boston, that same incident happened, they said. And they knew those kids were watching that movie, because otherwise where would they get it? I always said, where did the writer get it from? Where did he find this idea? Where did the writer of *Born Innocent* get his idea? Did he see it on television? Was it the first time it ever appeared? It can't be, he got it from something that's in the human being. Everything, every violent act, every stupid act, everything is within us all. We may be stimulated by something and react to it, or we may be stimulated and not react to it.

Starting Out

QUESTION: As a first-time producer just starting out, without any credentials, what would be the best way to present an idea that would be current in today's marketplace, which could create some excitement?

PAUL KLEIN: When I left NBC originally to handle a few people, a few things, I had some credentials, but not a lot. And if I had it to do over again I would hire an agent. That's the first thing. I'd give away a percentage immediately, because the agent has entrée into all the networks, and if he doesn't, then you get another agent, because they're easily changeable. And that's how you do it. I know you would like to present your own idea, but the agent sets you up to present it. He makes the appointment with the proper person, and then either he goes in and pitches it himself or you go in with him, and you pitch the idea, either in written form, or you talk it. I think you should talk the idea to the guy you're presenting it to. And you may make a sale, who the hell knows. It's possible to make a sale your first time out. Unlikely, but it's possible. Networking is a business, it's a profession, it's a business like a store, like Bloomingdale's or I. Magnin. And we buy from suppliers, we don't manufacture everything ourselves. We manufacture virtually nothing ourselves.

We think we know what's selling. We're buyers, we're merchandisers, we buy product from other people. In order to buy from you, if I were a store and you were selling suits, I'd want to know if you can make these suits. Can you deliver? You know, there is no novel idea. Aeschylus had the last novel idea. It's very, very difficult to come up with anything new. You've got to take what's old and make it new, and manufacture it, and have it packaged well so that we can sell it. We distribute it, and people will then buy it based on the package and whatever glitter we can give it. So we would be hesitant to buy from you as a new producer. Because you're not a producer. You say you're a producer. I could say I'm a pants maker. I'm desperate, I want to go into that business. I go to the buyer at Bloomingdale's and they listen to me, and then they say, "I don't know, did you ever make pants before?" I say "No, but I really know how to make it. And I have this terrific design with things on the bottom."

But it is very difficult to get in. We want to buy from professionals. Sometimes we are convinced to buy from an amateur, which you are if you've never sold anything before. You become a professional after you've performed the first act, I think. For money.

SONNY FOX: But there is a way, let me hasten to add, because I did this a couple of times when I was at the network, and we've all done it. Somebody comes in with a good idea, and you think it's a good idea. And you say to the guy, "OK, I like the idea, but I don't trust your ability to deliver it. Find somebody that's acceptable to us as a bona fide producer and associate yourself with him and then come in together, and then maybe we'll make a deal." That guy then becomes the guarantor, in effect, of delivery. I wouldn't tell the guy who to get, and I don't say you must go with somebody.

We also do that for financial reasons. A guy will come in and we have to have reasonable assurance that economically he will complete the performance and the delivery, and if he goes into deficit, which producers often will do, that he will not suddenly say, "I don't have any money, I can't finish it." So the business affairs people will insist that you have some kind of guarantor. It could be a bank, or it could be somebody like a Chuck Fries or a Universal, who will take you in and say, "Yes, we will back this man."

Now we're at the end of our discussion. I know that hearing from someone like Paul Klein is really like getting an inside look at television which is rarely given to people on the outside. One of the things that you don't know from Paul's very relaxed discourse, one of the things that you don't know unless you're sitting in the seat he occupies, is the tremendous pressure that converges on that one point, the pressure that emanates from above him and from around him, outside the company. It is a job which very very few people could even begin to handle, and until you're inside the network as I was recently, you really have no idea, even though you've been in the business as long as I have, of the pulls and the pushes on an individual. And there is only one individual finally that it comes down to, who controls so much money, so much power. And yet he has his bosses that he answers to, and he has his job on the line and his neck out there every time. So although we haven't gotten into a lot of that, and Paul's relaxed mood would belie that, it is indeed a truth and one of the factors that everybody has to live with in the network.

I certainly hope that you have gathered one salient fact, that commercial television is primarily a marketing medium, and secondarily an entertaining medium, but it is primarily a marketing medium, and the figures that Paul was using, which we would call demographics, the figures in terms of the population breakdown and the habits and the patterns of viewing, are the things that really are what shapes a network, and the reason that all three networks look more like each other than unlike—because the parameters that they deal with are the same, and whether Fred Silverman works at ABC or CBS or NBC, Fred Silverman still has to work with those same realities. Paul works with those realities, and it is not enough to beat the networks on the side of the head because they are not what you would like them to be, they are what they are because they have been formed by those factors, and some do better and some do worse, and some dream dreams and some don't, but the realities are there. The maneuvering room is very slim, and the stakes are very very high.

The Independent Producer

Alan Landsburg

Alan Landsburg is President of Alan Landsburg Productions, an independent production company he formed in 1970 to create television specials and series, feature films and movies for television. A native New Yorker and graduate of New York University, he began his career as a specialist in News and Public Affairs programming at NBC in New York. After 6 years at NBC and CBS he was named Producer-Director of the *Biography* series for Wolper Productions. In the years before he had started his own company he produced the National Geographic Society specials, *The Undersea World of Jacques Cousteau* and several dramatic specials, including an Emmy Award-winning Hallmark Hall of Fame presentation in 1970. Alan Landsburg Productions has furnished a steady flow of varied programming to all three networks, including the highly successful *In Search Of* series and the critically acclaimed *Between the Wars* programs. Landsburg has also been author or co-author of several books relating to his television productions.

Let me explain how an independent producer comes into being. Of course, there's really no such animal—I'm hardly independent. Were it not for networks and buyers I wouldn't exist. There are, in fact, very few successful independent producers currently surviving in the television industry. TAT, that is the Tandem Company, owned by Norman Lear and Bud Yorkin, qualifies as a successful independent production company. MTM, the Mary Tyler Moore organization, qualifies as independent. They own no studio. They don't have to produce shows to support a large physical investment or a production studio, and that is the definition of an independent. Warner Brothers must produce in order to survive because they have that bloody lot sitting out there. Columbia, Paramount, Twentieth Century-Fox all have lots and overhead on those lots. Therefore, they must produce to survive. An independent has no obligation to produce except in order to do his business. He doesn't sustain anything except his rent, his telephone, and his typewriter, and if he buys his typewriter he's crazy. He should rent it, because he can be out of business in a minute. We always say it's like the old merchant attitude—you have to be able to fold your tents and get out. If you don't, you die. I've seen too many companies go that route. Companies which stayed around long after the need has passed for their product litter the history of television.

Getting Started

I didn't start out in this business as a producer. I began as a writer and a director in radio. There wasn't any such thing as broadly applicable commercial television at that time. When I started at school there was a Department of Radio, and as I graduated from college it changed its name to the Department of Radio and Television, and nobody knew what the hell television was. I became a writer and not an entrepreneur. I was part of what was later to be known to me as the "creative community." I only knew I was writing and struggling for a living. Writing carried me through all the junior years, until I had written just about everything I could write on New York radio, and then got a chance to move to California and write Hollywood television, thanks to David Wolper, who was forming a new company and needed underpaid people to work 20 hours a day. And there you have the essence of what an independent television producer is. He's someone who manages to squeak by in the current economics by underpaying most of his staff.

When I came to California I worked for David Wolper, along with a handful of people who became the nucleus of that organization in the city. When I began at Wolper my contemporaries included Jack Haley Jr., who now runs his own company and produces Hollywood movie shows, now represented principally on the air by a series called *That's Hollywood* which he does in association with Twentieth Century-Fox. Jackie, in fact, became president of Fox Television for a while after he left Wolper. There were also Warren Bush, who runs Warren Bush Productions, Bob Guenette, who runs Guenette-Asselin Productions, William Friedkin, whose name I'm sure is familiar to you through his directing assignments, David Seltzer, author of *The Omen* and now head of his own production company, Mick Noxon and Irwin Rosten (now known as Ronox, responsible for the National Geographic specials) and Mel Stuart, who has joined my company.

That's half of them, a quick list off the top of my head. I know there are others, and each of them ran their own companies as independent producers because they came out of a stable producing environment, which allowed them to become known to the buying community. Wolper Productions was kind of *The New Yorker* of television. It poured in a lot of talented people and it poured them out just as quickly. I don't know how. I think the magic belonged to David himself, because David was a genius at picking people to do jobs. Everybody he picked has gone on, verifiably, to run their own companies, and they learned how to run their own companies from David. He picked the people, nurtured them, and when it was time for them to move on, let them move on. I started my own company when I left Wolper in 1970, and I've been running my own company ever since.

I think this kind of background helps to explain my own success, and certainly longevity, as an independent producer. I have managed to survive in roughly the same capacity for 18 years. People with whom I have dealt on a daily or monthly basis have by now reached the zenith of their individual and collective careers. After 18 years of struggling we are still talking to each other, no longer at the level of starters, but now at the level of the "in" group. I no longer have, among these peers, to prove anything. Old friendships thus provide an atmosphere for me to walk in and sell my merchandise to the people who are buying.

The economics of running a television production company are very strange. Only in television is a product sold before it is in existence. I don't know of another business in which that happens. We dream up an idea, walk in to a potential buyer and say, "I have this idea." It's a very gossamer thing. We have in our office about 14 file cabinets where we have written ideas in some form or another. That amounts to 2,000 or 3,000 ideas, and time and again I am called upon to walk into an office, to sit with a man and say, "I have an idea."

The whole process of building a catalogue of ideas is one that I think is valuable. Don't lose an idea because it's hard to get it accepted, and don't be momentary with it, because good ideas will come around again and again and again. The Tandem Company, which has been a major force in programming for several years now, all grew from one series idea. Norman Lear and Bud Yorkin had been part of the television industry as comedy writers for fifteen years before *All in the Family*. It took them 5 years to get *All in the Family* on the air, and even then it was only freaky that they got it on. Their talent and their experience allowed them to understand the process by which a program grows, and it is not a simple process. Their understanding of all the talents that go into making a program is what made that company work. It's what spawns all of the independent production companies, not only Norman Lear's. It's all that experience.

Selling the Idea

The only reason that my idea is given a hearing is because somewhere in my past I have exhibited the capability of not only describing that idea, but realizing it in television form. It's a very strange process, and let me walk you through every step of it as I have seen it.

There are, within each network, seven major departments. From my point of view, I look to the program departments of a network as my market. Within that program department, there are generally the low-scale divisions, the divisions that are the most profitable, that get the least attention and are subjected to the least amount of pressure, in relative terms. Those are called day part and late night — daytime programming and late night programming, and included within day part or daytime is children's programming. Those are three markets, three heads of areas with whom I must deal.

There is, after that, the department handling specials. In today's fragmented television world, the man who heads specials walks a very thin tightrope, because almost everything everybody else does becomes a special if it is nothing else. So the man heading the department responsible for producing specials quite often doesn't know when he's going to get air time, what air time he's going to have to program for and what will be considered a special when he programs it. There used to be clear divisions between series and specials.

There are then the straight-on series groups, and they

break into two divisions. One is always considered long-form drama, that is, one hour in length. It can be, and is often, by the shibboleth of the television industry, action adventure, medical, or legal. All of those shows fit into the one-hour series form. The half-hour form is only comedy, with a few notable exceptions. *Adam-12* was a half-hour form that was not comedy. But otherwise, half-hour form is situation comedy. Both long form series and situation comedy, or half-hour short form, are run by different department heads.

There is one more area we deal in quite a lot, and that is a new area in television called movies. All three networks have a separate department to which I bring specific ideas for movies for television.

For example, going for a two-hour movie, I will walk in and say, "I would like to do a movie involved with insects in a small town killing people. Probably bees. Do you want to do that kind of movie?" The answer is quite often, "Yes, that fits our program needs, but it should really have three attractive women, a lot of killing, but bury it at the end, and a lead man, because we have someone under contract now. Can you do a movie like that?" It has obviously changed my idea about a town surrounded by bees, but that's OK. We are obviously now meeting each other's needs. That is the way an idea begins. The business process which will then follow is for me to say, "I believe that Mr. Fox, a writer whom I have in mind to do this film, should do it," and I will then get agreement that, yes, if we're going to do those three ladies and the deaths at the end and the bees somewhere in the middle, it should be Mr. Fox. That is the last meeting at this stage of the game, and now it becomes a business affair.

In terms of business, I then send my business representative, in this case my partner, to open negotiations with the network. The idea has been as fragmented, I promise you, as I have just described it. Yet, from that idea, my partner and the business affairs division of the network must structure an entire production arrangement in steps.

That means that we now begin with a treatment. The first thing Mr. Fox will be called upon to do is a treatment, a detailed narrative description of the movie. Treatments are pegged at a minimum of $7,500. That's a Writers Guild minimum for a treatment. The network may choose, after seeing this treatment, to exercise a cut-off. They may decide to abandon that particular film at that point. But contractually, before we can sign to do that treatment, we must structure the entire business arrangement, which will now go through script first draft, script second draft, polish, production, cast approvals and cast budget, and delivery schedule. All of that must be contracted.

Once we have said yes, it would be a good idea to deal with three ladies, bees and a town, we don't know how much that film is going to cost. Neither does the network, but we must give a specific cost. We will negotiate it saying that, based on our experience, when you have bees and those ladies it's going to take $1.1 million to do this movie. And then they'll say no, bees, ladies and death is only $900,000. If you had costumes and fire it would be $1.1 million. And so we negotiate. The range currently for television movies is, low end, $925,000; high end, at the moment, about $1.3 million, and that's a figure that's really not locked in. It keeps on going up, maybe to $1.4 million. The range, though, is really $925,000 to $1.3 million. And this all has to be negotiated before anybody knows what the movie is about, or really cares.

Now the real work of creating begins, because the network, from its programming point of view, based on what has been successful for them, or for someone else, has made a determination—this is what they want. We now assign the writing of the treatment. There are an endless number of artistic permutations at this stage. Writers may have a treatment cut-off, a first draft cut-off, or a complete script. Now cut-offs are a very funny, ingrown process. If Mr. Fox had been the writer of seven successful movies, or even one, prior to our committing him to write this movie, he would not accept a cut-off. He would say, "Regardless of what you do with my work involved in this movie, you will have to pay me my going rate for a script and that is $35,000 minimum." For top writers, you can figure $35,000 for the script. Along the way Mr. Fox will grant us consultation, or even presentation of a treatment, but he may not be stopped at that point from earning the rest of his money. We may decide after looking at the treatment, that it is utter nonsense, that he is a boob and cannot write, that we are in a disaster zone. We still are obliged to pay him $35,000, and then we can bring in another writer and pile the money and the ideas on top of Mr. Fox's.

New young writers are forced, by the nature of the business, and by the nature of their desire to have their work progress, to take cut-offs. For that very reason, the second time you're faced with this deal, when you've had a successful movie, undoubtedly your reaction is "no cut-offs." First draft cut-offs have become, therefore, a kind of middle norm, a compromise. First draft cut-off says that we don't cut off the writer and his work at the treatment stage, but we have the right to take his first draft screenplay and say "thank you and goodbye." The first draft cut-off is generally a $20,000 arrangement, so that we will pay $20,000 and the writer will present us with a first draft script. At this point, I as producer have joined forces with the network, and the writer is alone out there trying to prove himself to both of us, and the network holds the controlling cards. When we get a first draft script and read it and find that everything seems to be in order, that, by God, we do have the bees, the three girls and the deaths, we can then say to the writer, "OK, we exercise the rest of our options on this and ask you to do a second draft for which we have committed an additional $15,000 to get you to your final price of $35,000." We also have the opportunity of saying goodbye, and turning to another writer and say "fix this."

In all of it, neither the writer, nor the network, nor my company, can determine whose final credit will appear on the screen. That the Writers Guild does. Once there are two creative hands, the Writers Guild takes all the material and saves it, weighs it, presents it to three of its members who determine who gets the credit. The credit is only important in that the residuals will flow along the lines of credits. So as that movie plays down the road, the only reason we ever go to credit arbitration is that the money flowing to the writer will be determined during the course of that action.

Now, that is a process that is both business and creative in nature. To say that any of us is simply creative or part of that creative community is foolish. The writer has been involved up to his tail in this business negotiation, has fashioned what he wants in a business sense before delving into the creative process. Because, after all, the creative process here, before he writes one word, has gone only so far as to say, "We want the following ingredients." There are nine people to whom movies may be sold directly, and only one is a woman, and she works for NBC. The great process by which we, as old friends, as part of the community, operate, is not the way that someone just starting would, and that's why I said to you the key to any success our company might have is survival. I've just been around long enough and have delivered enough material so that I can walk in with this sort of fragmentary approach. "I would like to do a movie about insects." Because my company has done movies called *The Savage Bees, Tarantula: Deadly Cargo, It Happened at Lakewood Manor*. These are all like the movie I just described to you. A lot of bugs and people dying in a town. Sure, I hope they're better than that. I want them to be better than that. And I work on them with the same passion it takes to work on something that has what would be called by my mother, proudly, quality. It doesn't matter. We work on them just as hard. The process by which we arrive at them is a little bit different.

At every stage of these negotiations and development in terms of the movie, the network money is at stake, not mine. What I supply them with free of charge is my time and my brain, at this point in the process.

As we complete all of the drafts of a script, we present them to the network, consult with them about the various pluses and minuses and worry about whatever supposedly questionable words are included in the dialogue. I've had long discussions over "bastard," whether or not "bastard" is appropriate in the 8 to 9 area, or whether it's not appropriate 8 to 9, but after 9 o'clock you can say the word despite the content. Those are niggling problems. There are people with the networks who are charged with propriety. Their view of propriety certainly does not meet mine. I'm sure it would not meet yours, but they have a view of propriety. My own funny story in that regard was Continuity Acceptance not wanting to pass the word *basta*. They didn't understand that it meant "stop" in Italian, and not "bastard." We had to explain what *basta* was, and then they weren't sure that they were wrong, because if they didn't understand what *basta* was, perhaps the audience would misunderstand and be offended. Standards and Practices have, in only a few cases, slowed down the production of a movie or changed the course of production of a movie or series. They have created many ulcers. They have never stopped anything.

At this level of acceptance, when we have delivered our first draft, second draft, and accumulated all our comments, there's a magic moment in the life of this would-be film. That magic moment occurs when a network executive says, "You have a go." That's being awarded the medal of honor in terms of television because it means that the network will now fulfill the rest of your contract. "You have a go" means you can now produce the movie for the price that was agreed upon when it was just a fragment of an idea.

The Budget

And now the real problems begin for the producer. We are dealing now with the current two-hour movie form, and I'd like to take you down through what happens to me next. For the moment I have ignored all other forms, because the story will flow the same way.

What you see on page 48 represents sample pages of a budget summary. Behind every single one of these lines, all 42 of them, there are 8 to 12 pages of budget detail. I'm going to explain them one by one because it's what I have to deal with every day. I tend to want to forget some of the items because they're so absurd, but they are there and they are real, and I think it will help explain to you where the money goes in a production net of upwards of a million dollars—what goes into the average two-hour television movie.

We create a budget by taking a script and breaking down

Form 1 (top left): ALAN LANDSBURG PRODUCTIONS PRODUCTION BUDGET — Page 1

ESTIMATED DAYS ON PICTURE

Trvl | Holi | Idle | L/Loc | D/Loc | 2d Unit | Total

TITLE: _____
PROD.# _____
BUDGET DATE: _____

ACCT.#	CLASSIFICATION
	ABOVE THE LINE COSTS
P2-001	Story & Development
P3-026	Producer & Supervision
P3-045	Direction
P4-065	Talent
	TOTAL ABOVE THE LINE
	BELOW THE LINE COSTS
PRODUCTION	
P6-095	Production Staff
P7-120	Camera
P7-140	Special Photography
P8-160	Production Sound
P8-167	Lighting
P9-190	Set Operations
10-215	Special Effects
10-230	Property
11-250	Set Dressing
12-270	Set Design & Const.
13-295	Wardrobe
13-315	Makeup & Hair
14-336	Livestock
14-355	Special Equipment
15-370	Transportation
16-400	Stage/On-Lot Prod Fax
16-415	Location Rents/Expenses
18-450	Film/Lab-Production
18-470	Tests
19-490	Payroll Benefits-Prod
	TOTAL PRODUCTION COSTS
POST PRODUCTION	
19-501	Editorial
20-520	Sound Effects
21-540	Music & Scoring Stage
22-570	Sound Dubbing
23-590	Film/Lab-Post-Prod
24-620	Titles/Opticals/Inserts
24-640	Payroll Benefits-Post-P
	TOTAL POST PROD COSTS
GENERAL COSTS	
25-701	Legal/Admin Expense
26-750	Insurance
26-771	Publicity
	TOTAL GENERAL COSTS
	TOTAL BELOW THE LINE
	TOTAL NEGATIVE COST
	OVERHEAD COST
	TOTAL PICTURE COST

Form 2 (top right): Page 2 — ABOVE THE LINE COSTS — STORY & DEVELOPMENT

Columns: MEN | DAYS | RATE | AMOUNT | TOTALS

001	Writer
002	Rewrite, Polish
003	Secretary
004	Rights Reserved
005	Abandoned Properties
006	2nd Run Residuals
007	Title Clearance
008	Research
009	Writers Expenses
010	Mimeo Scripts
011	Royalties
012	Payroll Taxes
013	Pension, H & W
014	Miscellaneous

Form 3 (bottom left): Page 14

Columns: MEN | DAYS | RATE | AMOUNT | TOTALS

336	Wranglers
337	AHA Man
338	Fees, License, Vet Exn
339	Rentals
	Horses/Cattle
	Wagons
	Saddles
	Animal Feed
340	Animal Trainer/Handler
341	Picture Animals
	Dogs
	Birds
	Other
342	Horse Trucks
343	Special Equip & Transp
344	Miscellaneous
	SPECIAL EQUIPMENT PACKAGE
355	Cinemobile
356	Package Grip/Electric
357	Package Generator/Elect
358	Other Package Complete
359	Miscellaneous

Form 4 (bottom right)

Columns: MEN | DAYS | RATE | AMOUNT | TOTALS

370	Transportation Gaffer
371	Camera/Sound Driver
372	Grip/Electric Driver
373	Prop/Wardrobe Driver
374	Honeywagon Driver
375	Chapman Crane Driver
376	Special Equip Driver
377	Other Drivers
378	Location Drivers
379	Airport P.U./Del
380	Camera/Sound Truck
381	Grip/Elec Truck
382	Other Trucks
383	Honey Wagon
384	Chapman Crane
385	Other Special Equip
	Boats, Tractor, Military Equip.
386	Car Rentals
387	Insert Car
388	Sub Contract Bus
389	Fuel & Service
390	Repairs & Maintenance
391	Miscellaneous

Sample pages from a budget summary form used by Alan Landsburg Productions. The entire form runs to 28 pages.

every item in that script, pricing that item, adding up all the figures. That's what the movie will probably cost if you're good and don't make any mistakes, and if it doesn't rain that Sunday, and if Monday is not too sunny, you'll make it on budget. There are so many variables to making it on budget that we could spend hours talking about. When I finish doing my budget, I may want to go back to the network and say "Listen, fellows, for a million dollars I'd rather not do this movie. Take it back, because I'm going to be in deficit $500,000 when I finish and that's too much of a deficit to eat." Out of the kindness of their hearts we may get to $1.2 million.

Back to the items in the budget. Budgets break into two areas: above the line and below the line. Above the line is basically the talent. Now, what falls into the talent area is the writer and any payments made for rights that underlie what the writer has been writing. In other words, if I buy the rights to a novel, that will flow into a budget under "writer." The royalty or licensing fee for the movie rights to a book comes in at a mean figure of $15,000, with no continuing interest for original source material. Of course, there are tremendous deviations from that price, depending on how hot a property we're talking about and how much you want to do it. Anything we have paid for writers fits into above-the-line. We call it "story and development" in the budget.

Item two: producer and supervision. It means the producer and his staff—secretary, associate producer, production assistants—will all fit under that item. Having given you the writer fees top end at $35,000, producer fees range from $15,000 to $50,000. That enormous range comes from the fact that some producers would prefer a percentage of net profits from the ultimate gain of the film. Others take their money all in front. Taken in front, it's $50,000; taken principally in back it's $15,000. That's your range.

You must remember that the figures I've quoted so far are just for the first run, and the network, when it buys the movie for $1 million, buys two runs. The writer and the director, therefore, must be paid residuals for that second run, and any additional runs after that. So that the $35,000 fee for the writer will then be plus a negotiable residual which is a minimum, I believe, of $7,500. I'm not sure of my second run minimums. It's one of those figures that flows into a budget without my ever having to worry about it or negotiate it. It's almost automatic.

The director is the next above-the-line item. The Directors Guild has set the minimum for a television movie for one run at $25,000. Two runs would be $37,500.

The final above-the-line item is talent, that is actors and extras. Low end for a $1 million two-hour television movie, for one run, for cast is $100,000. High end for a two-hour television movie for one run is $180,000. You will find that the range falls in between the two. There's no way around it. We've tried every way possible, with minimums being paid to 85% of the cast, and it's impossible to do a cast budget for less. A day player gets approximately $400, and a weekly player will come to about $800. A star averages $20,000 for one run. A magnum star, and I don't even want to start to define what that is—a television important name—can get up to $50,000 for one run for his 18 to 20 days on a movie.

The second run will almost always be done at scale, even if you're a star. You negotiate the fee for the first run, and generally agree to pay scale for the second. So that the figure will drop enormously and would be about $40,000-$50,000 on a $180,000 cast budget. It would be about the same for a $100,000 budget. Your rerun will be paid at scale.

If you shoot a two-hour television movie in more than 23 days, you are in the toilet. If you can shoot a television movie in 18 days, you are roughly in fat city, because we are now at the below-the-line part of the budget, which is all other costs—and those you figure on a daily, or even hourly, basis.

We assume here that per day of shooting, the low end is $20,000 per day below the line; the high end is $35,000. You fight like the devil for every $1,000 per day, but it's very hard to imagine a film being made today for less than $20,000 per day, when you're on an average 20-day schedule. It's hard to be over $35,000, because you would then just be excessive and silly. Your cost per day is determined by where you are and who you're dealing with.

Your first large group below the line is your camera department. Generally, your camera department consists of everyone associated with the recording of the image. That's sound, picture, lights, grips, and electricians. Grips are the men responsible for planting the camera, for putting it down in its given place. The grip department is always a hard one to imagine—why do you need a department? Well, if you put the camera on a dolly, somebody's got to push it, if you put it on a crane, somebody's got to lift it, put it on the crane and run that crane to the place where you want it. If it's on a tripod, somebody's got to haul that tripod to the place where it goes.

Your general camera department starts with the DP, Director of Photography, who is the highest paid man. I don't know a DP today who works for less than $2,000 a week. He's responsible for the camera department. Below him there will be a camera operator who comes in at around $1,000 a week. Below them are at least two assistant cameramen and one apprentice, all of them responsible for loading the film and keeping the camera. The assistant cameraman is the man who keeps his hand on the focus, and he's just as important as the DP when you're shooting. If he blows the focus, you blow the shot.

Generally you have a sound mixer and a boom operator or two. The sound mixer is the man who runs the locked tape recorder to your camera. The boom man holds the mike into the scene. He may have one or two assistants. Your electrician is the man who sets up the lights and plugs in the juice. The DP tells them what to do. The DP tells the electrician where, how much, and what kind of light, but the electrician's got to bring it there for him. In general, you have a chief electrician, his best boy, and three helpers before you're finished with a movie. The price ranges in the $800-$900 per week category for the chief of the department, plus a box fee. An electrician carries with him all his little tools. That's called his box, which generally runs about $300 a week. Without that box you've got to go rent all those pieces of equipment, so you want his box. That's one of the pieces of froufrou, the dozens of additional charges that creep in. The grip and grip box gets about the same as an electrician. His best boy gets about $600-$700 a week. You may have two, three, or four in the grip department, depending on how many cranes and dollies you use, how many moves of camera you want. If you're going to be doing a lot of helicopter shooting you need someone to load that camera into the helicopter. That is your basic camera department.

Now, on top of that you must add all of the rentals for all of the boxes and the cameras and the recorders and the microphones. In addition, every time a light blows, it's $40 a bulb, and you must pay for those as well. You hope you're starting with fresh new bulbs when you rent the equipment, but probably the lights you get just came off another guy's shoot, and he's made sure that the bulbs have lasted right down to the last one. In fact, if you're not careful, he's plugged in some of his old ones so that you hit your first nine lights and six of them work. That's $120 bucks to turn on the other three lights. Most equipment is rented when you're an independent producer. All of these figures change radically when you're on a studio lot, because then the studio determines how much must be paid to each of the people and how much must be paid for the equipment rental. Equipment rental is not expensive. Lights, camera, sound equipment only runs about $4,000-$5,000 a week for the entire complement. Unless, of course, you're going out to a remote location, where you have to bring your own power, which means bringing your own generator, which means bringing a teamster to drive the generator there and to operate it all day long. He gets $900 a week. But he deserves it—they all deserve it.

Your next major area is set dressing and props, and that can get very expensive. Your art director is responsible for designing the set. Even if I'm going to shoot a scene in the actual location in which it takes place, say shooting a school scene in a real school, I'd have to bring my art director in, so that he can make sure that when the cameraman comes in and needs something to mask a problem area, he will have it. And when I'm shooting on a stage or building my own set, of course he's the designer, the one who makes sure that the picture will have a proper look. If we move into an apartment, he may take all the furniture out, rent new furniture and design in the look.

Below him is my set dresser. Once the art director has decided what will be in the scene, my set dresser runs around town buying, renting, borrowing, stealing and charging a lot for everything he does. They will work hand in hand with the prop man. The prop man is responsible for seeing that the license plates on the cars do not represent real numbers, for which we could be sued. He's in charge from something as minor as that, to supplying diving suits when we're dealing with an underwater picture. If we need a gun, he brings the gun. If you need cigarettes for the ashtrays, he's got a box full of dirty cigarettes. You pay for his box, too. In his box is every trick you can imagine. Almost anything you can demand for an actor to hold in his hand, the prop man will have in his box. He, too, is about $1,000 a week, but that will include his box quite often.

Then obviously there's the wardrobe department, with a costume designer who either draws and orders the clothing or goes and buys it. Designers can go into a department store and, if it's a modern picture, simply buy the costumes off the rack. We try to keep them out of Hermes and convince them to go to Thrifty to buy costumes.

The costume designer will have a man's wardrobe and woman's wardrobe handler. If there are stars of magnum quality in the picture they will have their own special wardrobe handlers. If you have a big cast of extras in costume you will need still another one or two wardrobe handlers on the days when all the extras are working.

The next item is makeup and hair. For makeup you need one person for the star, a hairdresser for the woman. Makeup runs about $800 a week, hair about $600, plus their boxes. They bring their boxes full of goodies, blowers, pencils—as always, the boxes are very important.

Any and all animals go into the livestock category. If you're doing a western, or even need a dog on the set, or a goldfish, or a spider, you get a horse wrangler, a goldfish wrangler, or a spider wrangler. Generally, about $500 or $600 a week, and they're paid for the week, even if you only need the dog for one little scene.

If you're using children, you must have a social worker and a teacher supplied by the city but unionized under IA. It's a great job being teacher on set! Working on a picture with children is a whole other horror. Once you put children into a picture you're allowed to work them, during a school day, four hours—a cumulative 4 hours in an 8-hour period. On a non-school day you can work them 6 hours in an 8 hour period. You must give them a rest period every hour, and they must be turned over to the social worker who clocks them in and clocks them out. She teaches them, sits them down, and they must have a separate trailer for a schoolroom, no matter where you are.

We are now at transportation. We must have a transportation captain. He's a union teamster who makes sure that all the cars driven to and from any place in which we're working, even if it's a studio, are driven by teamsters. He gets $1,200 a week, the teamsters get $900, based on a 12-hour day. They must report to work a week early, and must leave a week after the shooting to make sure everything is wrapped up. A lot of unions do that. The first assistant director must have a week to prepare the film, and gets at least a week's severance pay, even if he works only one week, so he gets three weeks' pay for one. The camera department gets a week to prepare and gets two days to turn back their equipment. If you don't give them these two days, the day you finish shooting, they walk off the set and there's your camera and equipment and somebody's got to round it up, check it out, and get it back in. It's better if your camera people do it, because if they are not responsible for doing it, we find, as producers, that lenses, drive mechanisms, cables, tripods, even cameras can be missing unless they are kept in the charge of the camera department and turned in by that camera department. It's a wonderful business!

If you're going to construct sets, you simply have to have a set construction department under your scenic designer, who will budget for set construction.

Location rentals—if you're going to go into a school auditorium to shoot, you have to have somebody make a deal with the school administration to allow you to come in to use the building. If the school administrator is smart, he or she will charge you for a guard and two cleanup people, plus the use of the hall.

That takes care of most of the incidental expenses. The others that will come up are film raw stock, that you can estimate in any budget at $320 per 1,000 feet, and that will include the buying of negative, the lab work, and the work print. Raw stock is the original negative. When we're on a set and we roll the camera and somebody says "Cut," and we're not going to use the material, we don't print that, so that becomes just raw stock that we've shot and not used. An average movie takes about 80,000 feet of raw stock, 60,000 feet of which will be printed. And of the 60,000 feet, the two-hour movie is roughly 10,000 feet.

Payroll benefits make up an important budget item because you allow 24 per cent of your total salary budget for benefits, that is, unemployment insurance, workmen's compensation, union fringe benefits, and that can be very stiff. That comes to about $75,000-$80,000 per production.

And then, your film is shot. Hopefully 22 days later, maybe 18 days later, God willing, 16 days, because then everybody dances in the street and throws a party. Then it goes to post production which is the next major set of items. Your editorial department needs an editor, and here you don't get killed too badly. You need an editor and an assistant. Your editor gets, according to industry custom, about $800 a week. Allow 16 weeks to edit the film.

Sound effects and music includes adding sound effects, cleaning up your dialogue tracks, scoring your music, adding your music. You can budget as a lump sum $25,000 here, unless you go get someone like Johnny Williams, who composed *Star Wars*. John Williams will make your music budget go up considerably, since his price as a composer is very high, and his demands as an artist are very high. So that, if you have John Williams you'll probably end up with about $50,000 for music and sound effects, whereas, if you don't have him, you're looking at $25,000.

An answer print (or proof) on that movie will cost $10,000. After you have done all your work, and your negative has been matched to your work print, your film is all cut, and it's all finished, just to look at it all finished costs

$10,000. The corrected first answer print will cost another $5,000. So it's $15,000 that you've got to spend just to see your movie.

Then you have costs which you will incur, legal and administrative, insurance, publicity, and overhead. I don't know a producer who does not have a legal retainer. Depending on activity, he is in the neighborhood of $5,000 a month. In figuring expenses you take the number of months you've been working on a picture and allocate a portion of the legal expenses based on that figure. The publicity department or firm must have a still photographer on set for whom you pay. He is a union member and he must be there to take all the still pictures. Generally his salary is $800 a week for the run of the shooting schedule. A publicity firm will generally get a fee of about $8,000-$10,000 for all of their effort, if they are hired day one and fired when the picture goes on the air.

Then there is that wonderful, mystical overhead, the cost of running the company that produces the movie. It can be as big as it is wide. It all depends on how many people you have to support on a yearly basis in order to mount a movie in a 4-5 week period. Overhead in an independent studio is generally laid in at about 15 per cent of the gross cost of the movie. The way you arrive at overhead is to take your business expenses for the year, above and beyond what is allocated in the budgets of the individual projects, and spread it across the number of pictures, dividing it equally. The more production you have, theoretically, the less your overhead. Practically, it just goes up.

That's why the two-hour movie today costs over $900,000 and somewhere around $1.2 or $1.3 million.

Why Be a Producer?

Through all of this very expensive process of making a film, there is a wonderful, almost magical sense of community which develops between all of the very different people involved in the production. There's a story about Sam Peckinpah, who was directing a film and moving right along on schedule. The last thing they had to do was blow up a bridge, and one of the last shots before they actually blew the bridge was a close-up of the dynamite being tied on. Just hands tying dynamite. That's something we generally leave to a second unit director, just tell him to give me six close-ups of the hands tying the dynamite. But it was the last shot. Peckinpah made it last a day and a half, only because he didn't want to say, "That's a wrap, that's it, stop," because then the community ends. They all fall apart. The fun, the excitement, the wonder, the glamour, the dazzle, the love, the hate, the stomach hurts, all of it, is compressed into those 22 days of work. It's when everything comes together. It's what all the magic of being a producer and being a creator is all about.

For me it's stalking behind the director and watching him set up what he's going to do, understanding what he's going to do, and questioning him about what he may be doing that may be wrong. Good producers and good directors must fight all the time. They must fight because one must question the other's judgment, and that's what breaks up marriages. "Why did you do that?" "I don't know why—my stomach said to do it, and that's why I did it." And you can get into a lot of blows. The only time I've ever physically hit anybody since I was 18 was in anger with a director.

It's very hard, but it's very exciting, so that's why, as a human being, I like to do it. There's an enormous satisfaction finally, the day I walk into the laboratory to look at the first answer print, having seen the picture at least 150 times through the editing process, and endless more times when we screen it to make sure it's right. But still, there is something about sitting there. That's the moment. It's not when it goes to an audience. It's not when it goes on television. It's the first time all the pieces are there and it's all up on the screen, and it's all mine, good, bad, or indifferent. A child has been born, and after that it's all anticlimax. That's the moment that really makes it worthwhile. And knowing what ideas, what feelings, what senses, what mistakes, what triumphs are sitting up there that I will never be able to communicate to you or to anybody, except to say they are there, and I know that they came out of some depth, some feeling that I had for the movie.

The rest of the business is that now I have produced and delivered a film to a network, at a likely deficit, ranging from $100,000-$300,000. What the network has bought for its money is a license to run that movie two times within a two- or three-year period. After that I own it, my company owns it. In its lifetime, if it is distributed in any number of forms in any number of places, it can earn again its original license fee. So to figure the potential profit you start with about a million dollars in the pot, and subtract from that my deficit, and the cost of money, and whatever other expenses I incur in the course of doing business. And that's why I do movies.

That money will come in over a long period, and will come for the following reasons. If we've done a successful movie, we may be able to find a theatrical circuit in Europe, the Far East, Latin America or Australia which will buy the movie, not as a television movie, but to play in their theatres, because there is truly very little difference between a movie for theatrical release and a made-for-TV movie. There is no difference in the design of the movie. There is no difference in the quality of the acting. There is no difference in the quality of production. So it can find a theatrical home in other countries. That is a dream, that's the blue skies, that's the clouds you see when you go to sleep and feel real good. It rarely happens.

More likely, this film will find its home in foreign television sales and domestic syndicated sales. Foreign television sales are obvious—England, Germany, France. There are a number of companies who specialize in the distribution of these movies. They will pay right now $300,000 as a guarantee for the rights to syndicate these movies in the United States, that is, to play them in the 11:30 movie slots you see around, or to sell them to television stations around the country, for whatever needs and sell them to foreign television. That $300,000 must be laid against any residual costs that I am responsible for, to the writers, actors, and so on. Therefore, in our initial contracts, we set those residual fees at scale. So that we again probably have to pay $40,000-$45,000 for the actors, and another $15,000-$20,000 for the writer, director and anyone else who may have a residual interest. That narrows the economic base considerably, and narrows the motivation for doing movies. If we see $250,000 as the net profit, out of all of our effort in the lifetime of that movie, we have seen a lot. If we can do six movies a year, we can therefore do a million and a half dollars, in long term net.

The reason there are very few independent producers surviving is, if you look at what I have just said, the payout takes 8 years. I've got to survive for 8 years to get all that money. If I go to the banks to borrow to live until all that money comes in, I will probably eat up most of it in the course of 8 years. So right now I'm walking head and head with the banks and I'm making them a fortune. But we survive. And we keep making movies. And, occasionally, a movie like *The Savage Bees* gets picked up in European theatrical circles and starts to make its own little culty niche, and blue skies happen. The rest of the movies don't, but that one carries it. It's the only business where there is finally somewhere a bonanza. That's why economically we stay in it. If it were only for the economics, I think I would rather be a writer, and grind out six scripts a year. At top end of the money, I'd do very nicely.

There are other joys of being a producer. There is recognition. There is control of the hours of my own life. To some extent, there is control of the product, and that's why I do it.

What does it take to be a producer? Most of it is having a good idea, but having a good idea and not being able to execute it makes you not a producer, but a writer. So if you're going to start anywhere, start learning how to write that idea, because, in the end, there are few entrepreneurs who are also writers. Most of them are not, and I have always felt that I had an edge because I started as a writer. The writing was really the result of training. Being a producer really comes out of 5,000 years of Jewish tradition.

Being a writer and writing the idea is how it begins, and hopefully you can wave those papers long enough in enough places so that someone says, "Yes, we want that, we want to progress with that idea"—someone, somehow. The only way I know of to become part of the professional community is not to start as a producer, but to start in a position within a production flow, within any part of this production flow, in which the ideas you conceive of can find a presentation base. If they find a presentation base, then you will be ready to take the next step, that is, to fail the next nine times, and to fail to get any response for what you do. Then finally, on the tenth try, somebody actually lets you write a treatment, maybe, if you're lucky. Maybe on the hundredth try, who knows? You never know when lightning strikes. It's the lightning strike you're looking for. The only hope is to be in the place where that lightning can strike.

As you continue to work in this business you get a feeling for where you should take a particular idea, where you are most likely to get a positive reaction. That involves the networks, what they feel is the right kind of program for what they conceive of as their audience, and it also means knowing what will appeal to the taste of the individuals in charge of the different areas at the networks. I can cite the pretension each network has, and its rationale for selecting the material, but that rarely has anything to do with the intuitive feeling, what Rommel called "Fingerspitzengefühl," that feeling at the tips of your fingers—"this is where I'm going to go. It's going to work." It comes out of recognition of what these people are. I have dealt with them, drunk with them, lunched with them, we have shared life experiences over glasses of wine. I know if I am dealing with a person who has had these life experiences, and I come in with something that is anathema to them, they're not going to respond. But I can try to relate it to what I know about them, knowing, for example, that one of the network movie heads is a film buff. He's a real dyed-in-the-wool buff who has seen almost every film, and so have I. We share being in love with film. If I want to relate an idea to him, I relate it in the terms of a film we have both admired, the kind of texture that that film had. It's a shorthand that allows me to broach an idea in thirty seconds. I can say "It's like Welles in *Touch of Evil*. It has that scary forcefulness of the long, honest shot." "Yes, let's do it." I've related to the person at a level that goes far beyond anything that's real in terms of business. It is really a matter of picking the person to whom your ideas and your personality and your image, if you will, work. It has taken years of work to get myself to that place.

SONNY FOX: You also have to know what else they're doing, it seems to me. For instance, if you had an idea for a program series that dealt with Germany or the years between the wars, you would not now go to NBC, knowing that they had a 12-hour series called *Holocaust* which follows some German families from 1935 on. So it's important to know what's on, what's in development and what's coming up. ABC is doing *Battlestar Galactica*, for instance, which is a great big science fiction thing. Anything else that comes in that has a science fiction label today is likely to get short shrift because of their heavy investment in *Galactica*.

The *In Search Of* Series

QUESTION: I was wondering about the *In Search Of* series, and what had to come about for you to get it on the air.

ALAN LANDSBURG: It's got an interesting history. Six years ago a friend of mine called and said, "I just saw a movie that you ought to be able to tear apart. I'm going to send the print over to you. It's a German movie, but you'd be interested in it."

I had nothing to do that night so I went in to see the German movie. His studio would not distribute it, but he thought that, just out of curiosity, I would like to see it. It was called *Chariots of the Gods*. It was a picture made in Germany by a German company. It was overlong, pretentiously narrated and absolutely fascinating. And I said it's all nonsense, it's full of garbage, there's no such thing as the phenomenon that they're outlining. I then wanted to find out the history of its distribution, and was told that there really was no possibility of getting it distributed in the U.S. in theatres. I happened to be working at that moment on the development of a series for Ford Motor Company. It was a very good, very solid documentary series, and it had taken me a year and a half to develop it and get it to Lee Iacocca, the head man, who finally said, "Yes, let's do it." I had a relationship with him, from having sat with him for an hour at a meeting and swapping stories about television, so that in 5 minutes he could say, 'Ford will sponsor the series."

So I called John Morrissey, his head of advertising, and said, "John, I found something that Lee would like. I'm going to send you a print, and I would like Ford to buy this. I'll cut it as a television special."

Four weeks later, John Morrissey called and said, "Close, but no cigar. We really don't want to touch it."

He called NBC and said, "I'd like to get this picture on as a Quaker Oats special," and they wanted Quaker Oats on the network enough to offer him the worst time period any special could ever be given—Friday night at ten o'clock, at which time NBC was an 18 share and death. He said, "I'll take it." Jack Young put up the money for me to buy the television and motion picture rights in the U.S. to *Chariots of the Gods*, which I took apart and re-edited entirely. I made a new picture out of someone's old one, like a retread.

I was about to send the print back when Jack Young, who had just taken over at Quaker Oats as head of advertising, a friend of mine (we'd met in New York when he was at Grey Advertising) said, "I'm coming out to the coast, and I need a special. Do you have anything?" I screened the picture for him, and he said, "We want to do that."

I said, "I hate the title *Chariots of the Gods*. It doesn't mean anything to an audience. I'm going to call it *In Search of Ancient Astronauts* because that's what it's really about, and if we ever can do it as a series, we will call it the *In Search Of* series." At which point my good friend Oscar Dystel, who is president of Bantam Books, called me in a rage because he had this little book out on the stands called *Chariots of the Gods* which had been a mild success, 30,000 to 40,000 copies sold at that point, and said, "How dare you do this to me! How dare you change the title!" I said, "Oscar, you can buy my title from me. Put a banner on it—'As seen on television as *In Search of Ancient Astronauts.*'"

Well, a lot of heavy language went down, and the movie went to television, and for some reason which I still cannot explain to you, since it was not heavily promoted, it did a 35 share in what had been a dead time period. It wiped out

CBS and ABC. Everybody was shocked. Jack Young walked around saying, "I'm a genius." I said I was a genius. Everybody associated with it suddenly overnight became a genius.

Interestingly enough, there was a little item in the film that said "Based on the book *Chariots of the Gods*." The next day there were over 300,000 books on order from bookstores around the country. Subsequently the book sold 5 million copies. One television show.

Jack Young didn't want to press his luck, so he refused to buy the next edition, which was *In Search of Ancient Mysteries*. We decided we would just go right on, change "astronauts" to "mysteries," and make another special, that Timex bought. Timex did equally well. Now, I said I'll be damned if I'm going to sell 5 million books for someone else, so *In Search of Ancient Mysteries* became a book that I wrote for Bantam, and a television special for Timex, and the book has since sold 1.2 million copies.

After *In Search of Ancient Astronauts* was such a big success some people came to us and wanted to buy *Chariots of the Gods* to distribute in theatres, because they still thought there was some juice in it. They bought it, distributed it, and it made something like $22 million in theatrical distribution. These same people came to me later and said "Can we do anything with *In Search of Ancient Mysteries*?" I said sure. We added 40 new minutes on top of the 50 old minutes, and it became the movie entitled *The Outer Space Connection*. That grossed $15 million in theatres, after being shown on television as *In Search of Ancient Mysteries*. But because it was called *The Outer Space Connection* Bantam did another book, which has since sold 700,000 copies. With that as a background, it was very easy to sell a series called *In Search Of*. It took no genius to sell it. It's lightning in a bottle. It happens every once in a while.

Producing for PBS

QUESTION: Can you make a comment about the 4th network, PBS (Public Broadcasting System)?

ALAN LANDSBURG: I don't really feel highly qualified to answer that. I have worked with PBS in the program development routine in the past, but I'm used to action now. PBS is a long process. For anyone starting out, I would say PBS is wonderful. It makes you go through all of the rituals that are involved in producing. You must budget accurately, to the penny. You must write, you must develop all of your ideas before you can do them. Within PBS, an idea must be literally a book of information that can be distributed to all the funding sources before it can become a reality.

Therefore, you are producing it on paper before you are actually producing. It's all drudgery, but it's all practice for what you want to do. I really fancy myself a Las Vegas gambler compared to that route, because all I have to do is walk in and say, "I want to do this and that," and I get to at least start it. I don't have to go through this whole process on paper before I do it.

Now PBS as an outlet for valuable work was always a kind of dream situation. If I wanted to do something that I felt was too narrow in range for network television and its need to capture a wide audience—and that's the key, really, to network television—I would bring it instead to PBS. If I wanted to do "Mozart, the Quest for the Man," to bring that to network television would be silly—it's never going to go anywhere. But PBS has a real need to fill the gap in the community.

My only distress with PBS is my recognition of their new seeming direction. They are now saying, "Let's popularize ourselves. Let's reach out and be a pop source to the audience. Let's get the ratings. Let's be more commercial." They are getting away from what I think they do best, which is to provide an alternative that is exciting and engrossing as an availability.

The PBS production process is, I think, too cumbersome, but necessarily so because it's government underwritten, not privately underwritten primarily. The new series I'm doing, *Between the Wars*, originated as an idea that might go to PBS, and it's the first time I know of that an idea originated with PBS, and for PBS, and went instead onto commercial television stations. But I had Mobil Oil Corporation. Mobil wanted to do it. Mobil was the underwriting source, and Mobil is one of the few companies, perhaps the only company today, that has the dollar power to say, "I'm going to invest my $3 million in the cost of putting this program on the air." Not the cost of the program—the cost of putting it on the air. Lay on top of that the cost of program, and each half hour is enormously expensive. It will, however, gather honors in a remarkable way. And it will wind up, after its two commercial runs, running forever on PBS.

Film versus Videotape

QUESTION: I know you have a love and affinity for the quality of film, and I know a lot of that film work is disappearing in television production, and being replaced by videotape. Is this really a trend, and how do you feel about it?

ALAN LANDSBURG: There is a texture in film that is not available in the electronic medium, which is tape. And I think you're asking for the difference in quality and feel between electronic and film work.

I don't know if film is disappearing, if only because it has a lot of aging practitioners like me, who understand and deal comfortably with the electronic process, but love film. It's just a subjective, tactile sensation. I can touch film. I can't touch the image on tape. Using tape removes me just another arm's length. There's no reason why tape shouldn't take over. It's more economical, better, lighter, faster, probably equal in quality. But my ability to deal with tape in the same creative way as I deal with film is limited. I feel it's limited. I've talked myself into that. I understand everything that will happen to a piece of film, and I'm a little hostile to and suspicious of the computerized numbering system which finally produces married imagery and sound in tape. It's so purely subjective, but it's so much a part of what the creative process is anyway, that while I'm around and my generation of filmmakers is around we will continue to shun the use of tape. We are filmmakers because we love the pretension and the feeling of that word, that form, that medium. I don't get off on tape. Sitting in a dark cave, watching an electronic screen, and editing by cutting at the next number and gathering the pieces together on tape is no thrill for me. It is for people who have been raised with it. There is a generation coming, that is most of the university trained people, who are now getting into places where there is available tape equipment, and they are training on tape. They will come to love tape as much as I love film. Film will disappear as this decade's group of film people pass out of power places in the industry. So I think it is passing, but I'm going to be gone when it does pass.

SONNY FOX: One specific place where it is passing out is in the 3- and 4-camera shows that are shot in studios, usually sit-coms, where by slaving each of those cameras to a tape machine they can then post edit in tape. But that's under very controlled situations in the studio, where the very crisp feel of tape is OK. If you're going to shoot a movie on location, I defy you to do that show on tape. The state of the art, forgetting for the moment the predilection of the producer, is not ready for the kind of modeling lighting, the kind of softness and focus you can get with film. But if you're in a studio with controlled light circumstances, where you're doing a kind of play on film, it works fine, and pretty soon all of the sit-coms will probably be done on tape.

Pay Television

QUESTION: What do you think of the movement towards pay television, and how do you think it will affect the production process?

ALAN LANDSBURG: It's going to have the effect of adding 60 networks to the existing networks. That is, pay television now complements all of the free broadcast means. It simply fragments the total television audience into so many smaller segments. Because we are a nation of diverse tastes, there will now be 60 more opportunities to choose programs instead of 2 or 3. As soon as you fractionate the audience into 60 little pockets what you will see will be of lesser quality because there is no sustained and concentrated economic force. So I think the initial impact of broad pay television will be to lessen the quality of any individual hour and minute, but to give a much broader range of available information.

We have to solve and change all the economics of film and tape in order to have effective pay television. But we will, we'll change all the economics. Right now, our economic patterns are so bizarre, we are forced into such strange payment schedules, we have accelerated the cost of movies to such an incredible extent, that what we are doing is putting theatres out of business. There is no difference today between the television movie and the theatrical movie in terms of budget, quality of performance, or quality of ideas. And we're even finally starting to be able to say the words and manufacture the ideas in television which had formerly been reserved for theatrical films. No one can tell me the differences between the two any more. I just know there's a different way to capitalize on them. But there's no difference in the program material. Sure, you can hold a longer shot and a wider angle on a big screen when you don't have commercial interruption, but there's no real artistic thrust under it that makes the films different. So pay TV, I think, until we adjust the economic structure, is going to have the effect of doing to television what television has done to motion picture theatres, and that is to narrow the range and quality of the product.

Syndication

Norman Horowitz

In April, 1978, when Norman Horowitz made the remarks which follow, he was serving as Senior Vice President in charge of worldwide distribution for Columbia Pictures Television Distribution. On June 5 of the same year he was named president of that organization.

Horowitz first joined Columbia Pictures Television in 1956, became director of operations for the company's international production and distribution division in 1960, later became executive assistant to the President of Columbia Pictures Television International, and was named a Vice President in 1967. In 1968 he left to join CBS, where he served as director of international sales for CBS Enterprises. In 1970 he returned to Columbia Pictures Television as Vice President and General Manager of Columbia Pictures Television International, a position he held until he was appointed Senior Vice President in charge of worldwide distribution, assuming responsibility for all the division's syndicated sales activities.

I think it would be reasonable to start out by saying that there is a basic misnomer in referring to my work as "distribution." Perhaps it is appropriate to discuss the distribution of coffee, the distribution of automobiles, the distribution of cereals and things of that sort. I guess if we're looking for a brief description of what we do as a business, it would be appropriate to describe it as my Air Force roommate did a couple of months ago. He came to my office, and for about 30 minutes I described what I do for a living. Finally he looked up at me and said, "You're in the used film business."

I guess basically what we're talking about here is the used film business. It is a huge job of selling and distributing massive amounts of film throughout the world. Our job is taking a spool of film, that might be a print of *Gone with the Wind*, sending it to a television network in Japan, and having them send us $2 million. That little spool of film can be worth $2 million, or $200 if it's sold in a smaller country.

The dollar volume is hard to pin down. The business of syndication, or the used film business, will swing up and down so wildly as to make a specific statistic for a particular year a meaningless number. There was a lot of publicity in 1978 when Warner Brothers announced that their syndication business was $140 million, which was probably the largest year in the history of Warner Brothers in the used film business. Columbia's annual gross, I would say, will vary from approximately $50 million in 1978, up to possibly $150 million in the next year or two, since we expect to be distributing a major program in the U.S. shortly.

Internationally in the business of distribution, we employ approximately 130 people. Domestically, in U.S. distribution, about 25. There are, in addition, those faceless, nameless people who send out bills from the accounting department, who do our advertising and publicity, who ship our prints, who inspect our prints, who handle the technical aspects—probably another 100 people directly involved in that business. As far as people are concerned, it's not really a very large business.

The biggest company in the syndication business today is MCA, the giant of the business in all aspects. The other major syndication companies today would be Warner Brothers, Paramount, Viacom (which is the former CBS company), World Vision (which is the former ABC company), and Columbia Pictures Television. That basically would be it, since MGM is no longer a factor in the business.

The FCC and Independent Syndication

SONNY FOX: You mentioned Viacom and World Vision, companies which formerly were run by the networks. Why are the networks no longer in the business of distributing their own product, the shows and series that were on their own networks? Why don't they take advantage of this lucrative market as they once did?

NORMAN HOROWITZ: I spent 2 1/2 years at CBS in their distribution division, which was called CBS Enterprises. That economic honeymoon for the networks ended in about 1970. Primarily at the urging of McGannon of Westinghouse, the FCC passed its prime time access rule, one of the most incredibly stupid rules that has ever been promulgated in broadcasting, wherein the networks were forced out of the syndication business, which I think was appropriate, but for the wrong reasons.

SONNY FOX: And at that point they had to spin off their syndication businesses, so CBS Enterprises became

Viacom, and so on. Just quickly, the reason they were forced out is that the FCC felt the situation as it existed was restraint of trade. If I wanted to sell a series to NBC, for instance, NBC would say, "We will buy your series, but as a necessary predicate to buying the series, you must also sell us, in advance, the syndication rights." Therefore, if I had a hit show, I couldn't go out on the open market and get somebody to bid for it and really make some money on it. I had to give that away up front. With all the muscle the networks had, it was pretty tough to stand up to them and say, "No, you can't have it." If they said, "Goodbye, Charlie," then where did you go? There are only three ballparks in town. After a lot of complaints about that, the FCC said that the networks had to give up that whole area. They are now not allowed to go to a producer and syndicate his program, whether he wants them to or not.

NORMAN HOROWITZ: I'd like to make one correction on Sonny's statement about the networks. He said they made you sell them the syndication rights. That's not so—they made you *give* them the syndication rights to the programs that went on the air, and it really was an example of the exercise of naked power. If you wanted your program on the air, then you had to give the network the syndication rights, worth many, many millions of dollars. I have a lot of hostility in me for the way they exercised that power.

Columbia's Distribution Properties

SONNY FOX: Does Columbia sell anything other than shows which have finished their network runs?

NORMAN HOROWITZ: We have an enormous feature library which makes up the bulk of our business. We have acquired programming from independent producers, shows like *Barney Miller, Fish,* both from Danny Arnold, *Barnaby Jones* from Quinn Martin. We have cartoons also, most of which were produced by Hanna-Barbera years ago under a joint venture we had. That includes *The Flintstones, Huckleberry Hound, Yogi Bear, Quick Draw McGraw,* things of that sort. We also distribute ABC film and documentaries abroad.

We are also producers of original programming for local stations. We have been particularly unsuccessful in this regard. We have done a program called *Special Edition*, with Barbara Feldon, which has been seen by probably four people in the world. It plays on about nine stations. This is an example, in my view, of how all stations lie. They say, "Hey, give us something other than those crappy game shows, and we'll buy them." They lie. It reads well when they say it, and they sound terrific, but they don't buy it. We did a pilot for our own game show called *The New Quiz Kids*, which was a reasonable attempt at valid programming, but again, there's nobody out there to buy that kind of stuff. We have not really gone into the *$128,000 Question, The Price is Right, The Gong Show* type of program. We distribute basically anything we can get our hands on that we feel we can make a buck on while retaining a reasonable amount of dignity. By that I mean we've turned down the body-building shows, where they have great looking guys and gals flexing for the cameras.

SONNY FOX: Would you distribute *The Gong Show* if you could?

NORMAN HOROWITZ: I'd have to say yes.

SONNY FOX: I just wanted to see how far your dignity goes.

QUESTION: You've said your distribution company handles series and features. What happens when an independent producer has a property like a half hour special, say a *Peanuts* special or a Christmas or Easter show? Does that producer go to a major distribution company or directly to private stations?

NORMAN HOROWITZ: The answer relates to a basic problem of distribution. If one of our salesmen goes in to Pittsburgh and sells 268 episodes of the worst series ever made for ten runs over five years, that one sale generates an incredible amount of revenue from that particular market. Basically the same effort is made to sell one special to that same guy, so that from my point of view it is nearly economic suicide for a distributor to get involved with those single specials. To have one isn't terrible, but to have five or ten of them is asking to take a loss.

One guy came to me and said, "Look, I've devoted the last two years of my life to making this special and I want to make sure it's marketed well, so I'm going to work with your salesmen and help you distribute it." And I said, "Our normal distribution fee is 35 per cent, but if you help, it's 50 per cent, because in the flow of life, when you have 3,000 features, 200 series, all this activity going on all around the world, we don't need someone calling to find out what's happening with *Charlie Brown* in Denmark." So it's a very, very tough thing for the independent to get reasonable distribution on a special. But there are specialists in that kind of work, independent distributors.

Syndicating *Barney Miller*

SONNY FOX: To talk about how Columbia acquires a show for syndication and subsequently markets it, I suggested that we take one specific show which Columbia will be distributing, *Barney Miller*, which is a big hit show on ABC, as an example. When did you get involved with *Barney Miller?*

NORMAN HOROWITZ: The process began probably four years ago. Danny Arnold, a very talented independent producer, had a pilot at ABC called *Barney Miller*. He had a certain amount of money with which to make the pilot, and used that amount of money plus a lot of his own. The amount of money given to him by the network was inadequate to do the pilot the way he wanted to do it. He then succeeded in the network business by selling this pilot to ABC. Success in the network business is the ability to go and make more programs for the networks to lose a little bit more money, and there is ultimately a rationale for it. But in the beginning, it's a license to lose money, figuring the number of episodes you get a commitment for, times however much money you lose on each episode. Danny, on the level, hocked his house and other belongings in order to finance the pilot. He was then faced with deficit financing

13 half hour shows for a summer replacement on ABC. The chances of that show staying on the air were probably 50 to 1 against its staying on the air past its original 13-week commitment. Danny had a young man working for him by the name of Jordan Davis who took the pilot to the distribution companies—Warner Brothers, Paramount, Viacom, World Vision and the others—and said, "Hey, fellows, I have this pilot and I will give you an opportunity to have U.S. and foreign syndication rights on this program." There was one company willing to make the deal on the show. I saw the pilot and loved it. It has a very kind of New York ethnic appeal to it. I took some of Columbia's money—not very much, under $100,000—and said, "Here, we will buy the U.S. and foreign syndication rights from you," which we did, and that gamble on my company's part will now be worth a fantastic amount of money to us, and ultimately to Danny Arnold and his production company.

SONNY FOX: However, if Danny Arnold had been able to wait and not sell the rights before he started the production of the series, he would have made more money. Knowing the success of that series, if he had held on to the syndication rights, came to you today and said, "I want to make a deal"—what's the difference to Danny Arnold, do you think, between making the deal then and making it now?

NORMAN HOROWITZ: The total earnings for an episode of a successful series in the aftermarket today comes to between $200,000–$400,000 for each program. Let's assume that there will be a total of 100 *Barney Miller* episodes, and let's stay somewhere in the middle of the range, $300,000 a program, times 100 programs is $30 million. I would say he would have made an additional $5-6 million as his share of the profits by being in the position to have the unencumbered syndication rights to *Barney Miller* right now. But he needed money for production, and we were betting our money on an enormous longshot. It was a terrific deal for him, even if I had to tell him so.

We own the rights to the *Barney Miller* program, but we have to wait years before we can make any money out of that. On the basic network deal there's a five year network exclusivity. The networks buy a program and, while they continue to license that program, we cannot sell the old episodes for syndication. *Barney Miller* went on the air in 1975, so it will be available for syndicated broadcast in September 1980. It's now March 1978, and we are just getting ready to go out and start selling this show to the stations.

QUESTION: How much money do you have to bring in on the sale of *Barney Miller* for it to be considered a successful sale?

NORMAN HOROWITZ: It is very difficult to judge whether or not you have been successful in selling a show in syndication. There are no guidelines, no point of reference you can use to measure whether or not you've made the greatest possible profit in selling your product. It's not like selling coffee, where you can send out comparison shoppers and find out how much the other guy is getting for his coffee and you can price yours accordingly. There's a great fragility to the marketplace. When I give you a range of $200,000–$400,000 per program, I can't tell you beforehand where in that enormous range the sale of *Barney Miller* will fall. Nobody can, and nobody can judge whether the $300,000 we might get was the most we could have gotten for that program. The president of my company has said to me in this regard, "Norman, how do I know that you're doing a good job?" And I said, "You're stuck, you'll never know. You can read reports, call your friends, do whatever you want." There is no definitive way to judge the business of film and telefilm in a marketplace.

There are norms for programming that evolve in a market based on the dynamics of the marketplace, and what goes on at a particular moment in that marketplace. A market may be a hundred dollar market. What makes it a hundred dollar market? Three guys sat around, and the first guy said, "Look, I only pay $100 per half hour." The second guy said, "I only pay $100 for the same half hour." The third guy said $100. And by dint of the three guys having a basically illegal, silent conspiratorial agreement that the price is $100, it's a hundred dollar market. The advertising rate card can go up and up and up, and it's still a $100 market.

But that changes, sometimes on a moment's notice, and that's where the dynamics of the marketplace come in. Let's say that station number one has a new owner. Station number one then hires a new program director. The new program director says, "Screw everybody and their $100. I want to increase ratings. We're in third place. If I raise the ratings 3 points it will have a cumulative effect this year of $800,000, so I'm going to disturb the status quo." The next time the syndicators come in, he says, "I'll give you $150 for your programming," or "I'm going to have your programming, I don't care what I have to pay." That's really what affects the price I can get for my shows—personnel changes, new programming policies, new advertisers in the market, heaven only knows what else.

There are no reliable published reports of what people are paying for syndicated film in the U.S. You'll see it for overseas sales, because *Variety* prints an annual report on those prices. But we live in a world of secrecy as far as the prices people pay. They are afraid that if it is known that they paid $5,000 for a particular show, the next guy who comes in to make a deal will demand $6,000. The problem in assessing a value basically is that it's based on show business, on what kind of ratings the stations think they will get for the program you have to sell them. If you make an ashtray and it costs you 37¢ to produce you are going to base your price on that cost and not go too far beyond a certain percentage markup, because everybody knows about what an ashtray is supposed to cost. But here we are dealing with guesswork, with rating points, popularity, and the advertisers' desire to be connected with a certain type of show. That leads to tremendous arguments between the station people and the syndicators, because we're all basing our demands and our offers on our own judgments.

Guys have gone bananas for *Barney Miller*, they want the show on their stations, and they've already told me, "You guys are ridiculous with your prices," when we haven't even quoted a price yet. But they know we're going to kill them. I say to them, "You're absolutely right, we're going to want too much for *Barney Miller*. If you want to buy *Father Knows Best* we can talk about much lower prices."

We're going to do something interesting with *Barney Miller* in an attempt to get more money for the program. It would be reasonable to tell you that the domestic marketplace is pumped up by money from the independent stations. They are doing very well now, particularly in the year that ended in December 1977. This is partially due to a tremendous increase in spot ad business. There is a tremendous euphoria among the independents, and there is no better place to deal than a marketplace where the buyers are euphoric about the money that's coming in. They're experiencing 20, 25, 30 per cent revenue increases, upping their rate cards every minute and a half. When a station is in that position they say, "The money's coming in—buy, buy, buy!"

Dick Lawrence, head of domestic syndication for Paramount, sold *Happy Days* and really rang the bell. Basically he said to his associates, "Everybody wants *Happy Days*, and nobody ever killed the auctioneer. What we'll do is auction off *Happy Days*," and he did just that. He sent telegrams out to the stations saying, "We're offering 6 runs of *Happy Days*, over 3 years, payable as follows. The minimum price in your market is X number of dollars. We will receive bids." Well, he received bids, and they did landmark business. The gross in the first licensing period for *Happy Days* will probably be around $400,000, which is far and away the highest gross a program has ever received in U.S. syndication. You can compare that to a very successful *Mary Tyler Moore Show* syndication, which will probably do $150,000–$160,000.

What we are trying to do now—and I may get killed in the attempt—is to sell *Barney Miller* to more than one station in a city at the same time. What this idiotic plan entails is to go to the station and say, "Listen, by 1980 (probably sooner) *Mary Tyler Moore* will have run its course. We will sell you 520 runs of *Barney Miller* for broadcast at 4 o'clock. Your exclusivity will be in the day part, from sign-on until 6 o'clock." We'll then go to another station and say to them, "We'll sell you 520 runs of *Barney Miller* for 7:30 and your exclusivity will be from 7 o'clock till sign-off." This will put approximately 1000 runs into the marketplace, compressed into a two-year period, and this will accelerate the flow of the money, which is important in the business of television. It will give the program two bites of the apple, two chances for success, because if it work at 4 o'clock we'll have a 4 o'clock franchise, and if it works at 7:30 we'll have a 7:30 franchise. In this greed-motivated proposal, we would hope to make a whole lot more money from the syndication of *Barney Miller.*

Who Can Buy?

SONNY FOX: Any time you think you've reached the absolute limits of what the market will bear, somebody proves that you're wrong, and somehow or other, there's more money to be squeezed out of the market.

Now let's talk about whom you're selling these things to, the independent stations. An independent is a station that has no network affiliation, and therefore has to program its entire broadcast day, including prime time. Therefore they are a very good market for off-network syndicated programming. There are only about 60 or 70 independents in the U.S., out of 700 stations. Out of these potential buyers, how do you determine whom you sell a program to? When you go into a market, is it open to everybody? Do you simply say, "It's available, come and get it?" Do you go in and say, "Here's the price," and sell it to the first one who comes to you? How about if a group comes to you and says "I'll give you 3 stations," do they get primacy?

NORMAN HOROWITZ: Yes. The power play of television syndication is no different from power plays in all sorts of businesses. The big get bigger at the expense of the small, and the small yell like hell, and fight like tigers. It's very difficult to maintain your integrity, to deal in the marketplace and give everybody a reasonable opportunity to buy the program. The groups have tremendous power. That level of power rests with the guys like Metromedia and the other groups that own stations in multiple markets. They must operate, as far as the FCC is concerned, independently. They will say, "You don't have to sell me the program for Los Angeles, we'll consider it for Kansas City even if you don't sell to us in Los Angeles." But they lie, they lie like all powerful groups. It is really difficult to stay out of the clutches of the powerful.

SONNY FOX: As a matter of fact, there's a suit currently before the FCC in which a UHF station in Washington is suing Metromedia for being excluded from bidding on certain properties, including *M*A*S*H* I believe, saying it's unfair restraint of trade. They claim that they were prepared to top the offer made by WTTG, which is the Metromedia station in Washington, but they never got a chance because Twentieth Century-Fox made a deal with Metromedia in which they got New York, Washington, Los Angeles and a couple of other stations all at the same time. They were able to make one deal and get a lot of markets, which was better for them, but in the meantime the poor guy sitting at UHF who wants a chance at getting this product is closed out. That's an issue that is going to be fought out at the FCC over the next few years.

One of the other things that drives the local guys up the wall is that when they buy *Barney Miller* two years in advance, they have to start paying for it right away, don't they?

NORMAN HOROWITZ: That again is a matter of dynamics. When our people go into a market they're going to try to get as much money as they can for each program, and they're going to get that money as soon as possible. When we go in, we say to the station management, "We want $6,000 a run for your 6 or 10 runs—$3 1/2 million—and we want it all now." The station guy says, "Are you crazy—I won't have the program for 2 years." "All right—I'll be back tomorrow." The guy says, "Where're you going?" "Well, CBS is interested in buying the show and I think they're in a very strong cash position now, and they'll give us the $3 1/2 million."

Now if the station management really thinks you're going to get the money from CBS and if they really want that show, we end up with our money right away, or at least we work out an arrangement where we get a big chunk right away. Granted, this gives us the use of this guy's money for two years before he has the use of our program. Some of you will say this isn't fair. I have a personal thing that I do now in negotiations. A station guy says, "That's not fair," and I say, "Look, are we going to talk about fair? When did fair ever enter into this? To talk about fair we've got to go at this a whole different way. We've got to get somebody in to judge what's fair." A couple of months ago a guy from the regional NAB in Dallas who had a one-station market explained to me that the fair price for a program is the price of a 30-second spot. I said, "Wait a minute," and he said, "No, I've thought about it a lot, and that's fair." What could I say? The man has no competition in his little market. The day there are five stations competing in that market, that definition of fair is going to change.

Overseas Distribution

SONNY FOX: *Let's talk about overseas distribution, which is certainly a major aspect of your business.* Barney Miller *will have limited international distribution potential, will it not? Why? What is the nature of the international market?*

NORMAN HOROWITZ: Broadcasters all over the world relied on American programming when they first went on the air, that is in the Western countries. These are primarily stations that go on the air at 8 o'clock at night for 3 hours. They originally had one studio, a few cameras, a master control room, very few actors, unless they had an active film business, and they bought American programs. Basically they would buy anything you had. This was when the American television business was primarily a film business. Everything was on film. The original *I Love Lucy* episodes, *Bewitched, I Dream of Jeannie, The Donna Reed Show, My Three Sons,* and *Father Knows Best,* were all done on film. These countries developed their own electronic production capabilities. The Japanese, Spanish, Portuguese, and Germans, all went into electronic programming, but it was stuck in everyone's head that they looked to America, the land of Hollywood, tinsel, movies and glamor for film programming. They bought American film, but not American videotape. That's reason number one.

Reason number two is that the videotape programming in America is primarily comedy programming, reflective of the American idiom, the American lifestyle. American humor isn't funny to the Germans, the Spanish, the Argentinians or the Japanese; they each have their own humor. So *All in the Family*, even though it had its origins in a British comedy, is not funny to the television viewer in Buenos Aires, or in London. They look to America primarily for *Kojak, Wonder Woman, Police Woman,* or *Police Story,* for the film magic that is done here. We still try to sell most of the videotape comedies, but it's tough. There are complex union Guild restrictions that also make it difficult. The underlying reasons for the non-sale of videotape programming are the fact that the overseas marketplaces do their own electronic production, and the ethnic, cultural differences in comedy programs.

SONNY FOX: Now let's take the more successful overseas properties, like the film Westerns. How would you compare the revenues you'll make in American syndication versus the worldwide earnings?

NORMAN HOROWITZ: Very few Westerns are made today, so you can make very good deals. The market is escalating in a couple of key English-speaking territories. You could go into England today and say, "I have the only new Western series," and you'd get $25,000 a program. You'd probably get $15,000 in Australia. I'm making these figures up, you understand, but it will give you an approximation. If I went in with 22 episodes of *How the West Was Won* I'd do very well, an expectation of over $100,000 an hour overseas.

Overseas broadcast organizations are not like American networks in at least one respect. After the first 22 episodes most of them say, "The people are bored." Do they know the people are bored? No, but they decide they've had enough. So the expectation for each succeeding year of production will go down and down and down. If in America the 22 hours were to go on the air, be cancelled and go off the air, the possibility of American syndication is absolutely nil — the great American goose egg. So if all you have is that one year of programs to sell, those 22 hours, you have $100,000 or maybe more overseas and nothing at all in the U.S., but there is no future for the show after that one overseas run. If you get five years' worth, a hundred programs, of *Wonder Woman*, a kid-oriented, high-rated, afternoon programmable show, you could have a value in the U.S. for each one of the hundred episodes of maybe $300,000– $500,000. You're talking about $30, $40, $50, $60 million. As a syndicator you end up in the strange situation where the American failures are the overseas successes, and American successes are often overseas failures because of the inability of the foreign market to absorb the full run of 100, 120 or 150 episodes of a successful American series. They just won't play the complete run of episodes.

There are only a few international markets which account for the bulk of our foreign business. Canada is number one right now and exploding, because of the growing competition in the Canadian broadcast industry. Australia is very big now—again because of increased competition. The United Kingdom is also seeing increased competition for American programming. It's tempered by quota, but nonetheless it's there.

Most of the free money is picked up from those three major markets. What I mean by free money is no encumbrances of major taxation, major withholding taxes, and no major distribution expenses. You can get a lot of money in a Spanish language distribution deal, but you have to dub it, make prints, and send the prints out, which runs up a lot of extra expense. In addition, you're dealing in a lot of these countries with currency restrictions.

One of the things which makes foreign sales appealing to a large syndication company is that the international market exists parallel to the American network market. A program that goes on the air in the U.S. right now is available overseas right now. If you're a syndicator and you've put out your money up front for the right to sell this program, that market is your first chance to get back a piece of your investment. The international market may keep going after a program has gone off the air in the U.S., but the major money happens right away.

In fact, though, we place very little of our programming in

most of the markets out there, and there are reasons which go beyond the cultural differences I've already mentioned. I speak about overseas sales in terms of our successes, but there are innumerable failures. They bought *Soap*, but they didn't buy *Barney Miller*, they didn't buy *Fish*, they didn't buy all kinds of things.

There is an additional factor in overseas sales, a quota system which limits the amount of American programming which can be used. In England, for instance, there is an absolute quota of 14 per cent on American product. That means no more than 14 per cent of their broadcast hours can be made up of American programming. They want work for their people, and their unions are very strong, so the quota system was set up at the outset of broadcasting in England to keep them from being overwhelmed by American programming. Australia put in a similar quota system.

The quota in Canada is about 55 per cent, which is certainly better, but it's still a quota. Canada is a country tortured by communications problems. Their broadcasting industry tries to retain a Canadian identity, a nationalism, and it just isn't possible, because something like 87 percent of Canadian homes can be reached by American television signals. If you try to put on a program that demonstrates what it means to be a Canadian when *All in the Family* is on, you're going to run into great problems.

SONNY FOX: There are interesting ways around some of this. There are a number of co-productions done in Canada. For instance, *The $128,000 Question* was done up there for a while. I was involved in producing another show which was done partly in Canada, partly in England. If you do something in Canada with Canadian content, which according to the fairly complex schedule they have comes out mostly Canadian, you come in as a Canadian production. Not only do you get a better price for your program in Canada, because it's not charged off against your American quota, but you are now under a Commonwealth quota as opposed to an American quota when you move into England. Therefore you get a better price for that same show in England.

Acquiring Programs for Syndication

We've discussed what kind of arrangement Danny Arnold made with you a few years ago for *Barney Miller*. Let's say that I am an independent producer. I have just sold a series to the network, and I come to you, Norman, and I say, "I want you to buy the syndication rights on this. I have a 13-week commitment from the network." What kind of deal can I get from you?

NORMAN HOROWITZ: The kind of deal a producer now gets from a distribution company is very different from the way it used to be. Early on, when the motion picture companies had all the stars, all the talent, the writers and directors, and all the facilities, if you came to a movie studio and said, "I've got a terrific idea for a program," they took out a gun and said, "Put up your hands," and made a deal with you in which your survival was possible, but improbable. The basic structure of the deal made it all but impossible for the producer to make any money. This was in a period when the networks were not financing programming. The production companies, divisions of the motion picture companies, were doing the financing, and they called the shots.

The situation changed as the power of the movie studios was diffused. The networks took over financing production, so they turned around and did the same thing that the production companies had done. A producer would come in to a network with his terrific idea and convince the network to order the show. "And by the way," the network would say, "we want the distribution deal, but we won't charge what those greedy motion picture companies charge. We'll only charge 40 per cent." And again, the producer was given the privilege of paying out of the remaining 60 per cent all the distribution expenses. In the old days the print cost, the advertising cost, the publicity cost, and all the money for distribution expenses, in addition to all the residuals, came out of the producer's share. A distributor could decide to advertise every day in the trade papers, figuring that it wouldn't cost him anything—it cost the producer. They could take out $50,000 worth of advertising, and if the $50,000 generated $50,000 worth of business, the $50,000 expense was charged to the producer, and the distributor got to keep half of the $50,000 receipts. It cost the producer $25,000, the distributor made $25,000, and he had a terrific deal.

In any event, that's the way it was, but "it ain't that way no more," not when you look at the tremendous amount of money available in syndication. When an established producer comes in with a program now, I say to him, "Look, the jig is up." I put the guns away. He can go to Paramount, to World Vision or Viacom. He is a free agent and the network is paying for virtually the entire cost of production for his program. At the other end down the line are incredible profits if the show stays on the air and is syndicated. Even overseas the profits can be significant. The producer has power now. He can write his own deal as the situation exists right now, with Viacom, World Vision and the other distribution companies fighting over the hot properties, struggling to stay in business. You must realize the economics of this. Take a show like *Barney Miller, Happy Days* or *Laverne and Shirley.* If there are 150 negatives, and the program grosses $400,000 a negative in U.S. syndication, that is $60 million. If you take $60 million and assume a 10 per cent distribution fee, that's $6 million that you can get for that show, even on a 10 per cent deal, and that's a hell of a lot of money.

I guess the only guideline I could give you at the moment that would be totally accurate is that the whole structure of distribution deals has exploded, and there is no structure. Deals have gone from where they were about 10 years ago and before, where the distributor took 50 per cent, to today's deal in syndication, and I haven't made one of these yet, where the distributor gets 7 1/2 per cent. The producers are now in a much more powerful position, so they're shaping deals to suit their needs. Now instead of the syndication company taking the distribution expenses out of the producer's share, they're agreeing to take the costs off the top, before the split. Or the producer comes in and says, "Look, I will give you 40 per cent of the gross if you pay all the expenses out of your share. Do whatever you want. You want to advertise? Terrific—your money, you pay for it." The concepts are changing because the gun is now in someone else's hands.

The business of selling programs to syndicators is no different really than the business of a syndicator selling programs to a station. The dynamics of the business are really what will determine the deal that you're going to get. You can come in very naively and get totally screwed, because *Variety* doesn't publish the deals that are being made today for independently acquired programming. So if you come to a guy, he'll tell you how terrific you are, that you're marvelous, and that he'll give you the best deal. It's like going to a car salesman. The dynamics of the marketplace are as variable on the acquisition side of the distribution business, as they are on the station's side. Let's say you have a new boy in town. He may be the head of distribution for Paramount. His management has just given him a new 3-year deal. They pay him a lot of money. Again, the money involved is very, very large. He says to himself, look, they're expecting a lot. The standard deal now may be to give a producer $15,000 a program as a guarantee. He'll say, "I want to make sure this producer doesn't walk out the door, so I'll give him $30,000 a program." He wants to show his management that he's really in business. Suddenly, because of the dynamics that exist at that particular moment with that particular individual, the $15,000 norm becomes $30,000.

I was the victim of a similar situation myself recently. As of April, 1977, my company had done 11 pilots for the networks. Our success ratio was 11 divided by nothing, we had nothing on the air. We have a very large distribution company to support, and having nothing on the air won't do the job. At that point I saw the pilots for a show called *Soap*, which some of you may have watched Tuesday nights at 9:30 on ABC. These were the funniest two half-hour pilots I think I've ever seen. I went to the management of my company and I said, "Look, we've got nothing on the air. We really have to have this program to distribute now overseas," because contrary to what I said before, I felt that *Soap* had at least a reasonable shot as a videotape program in the overseas marketplace. I sent my management a cassette of the program, and the closing line of my message to my home office in New York said that I felt *Soap* would be "the biggest comedy success in the history of American television, unless it isn't."

The president of my company loved it and said to me, "Get the show." And I said, "I'll get the show if I can make a reasonable deal," and he said, "No, no, no, get the show." And there you had the setup for the dynamics of what I spoke about before. The producer knew he had a hot property, major and minor distribution companies were interested in acquiring the distribution rights, and I got killed. I got killed by making a deal that I was not happy with, and I'm still unhappy with it. But I made the deal in order to get the show, not based on mathematics and computing potential profits, but knowing the craziness of the business and the dynamics of the business, I wanted the program. I had the message from management who said, "Get the show." The last time he said, "Get the show," by the way, was the *Lorenzo and Henrietta Music Show.* We did not get the show, and the president and I had it out on that one, but I never heard from him after the show got cancelled after 5 weeks. Had the show succeeded, I would never have heard the end of it.

SONNY FOX: Stay with *Soap* for a minute. You made this deal for *Soap* at a time when you were not sure whether or not it would be picked up for another season. If it was not picked up for another season, you would have been in trouble with your investment in that show, wouldn't you?"

NORMAN HOROWITZ: No, not really, because what happens fortunately from time to time is you get rescued by circumstances. In this case people viewed the show overseas the same way we viewed the show here. The Canadians thought it was going to be the highest rated show on ABC this year, and they paid much more than they normally pay. It bailed me out a little bit. We practically created a revolution on Australian television with *Soap,* and they paid much more than they normally would. The British did not pay as much as we would have liked, but then again, substantially more than I thought they would. In that way we had a bail-out even if the show went for just one year.

Syndicating First-Run Programs

QUESTION: You've been talking a lot about used film. Lately, certain things like *Mary Hartman, Mary Hartman* and a number of other programs are not used film, they're new film and they go directly into syndication. What is the thinking behind doing that as opposed to going with the networks?

NORMAN HOROWITZ: *Mary Hartman, Mary Hartman* evolved, as a lot of those things evolve, from a very creative producer, Norman Lear, trying to do something for the network and the network saying, "What, are you crazy? We're not going to play it," and Lear not being willing to let it die. There are many, many attempts by production companies to do that type of programming. *Mary Hartman, Mary Hartman, The Lorenzo and Henrietta Music Show,* even *The $128,000 Question* are attempts by production companies to find a marketplace out there in the world away from the network dominance, but the marketplace is very limited because of the networks' stranglehold on the distribution process. Each network has the major stations in the major markets by the throat, saying, "You'd better play most of our schedule, or you'll lose that affiliation agreement." The producers who don't sell their show to a network but still want to sell a program for evening prime time (maybe a show like *Mary Hartman,* which could play at 9 o'clock at night) are then forced to go to the independent stations. Something like 60 per cent of the homes in the U.S. are within the reception area of at least one independent station, although a lot of those stations are UHF stations that deliver very small audiences.

These independent producers fall victim to the economics of the business by not being able to line up enough stations. You must keep in mind where the money is in television. Let's use *Mary Hartman* as an example—I'm going to make all of these numbers up, but you'll see my point. *Mary Hartman* has to be produced, and let's assume that it is produced for $130,000 a week—it was an inexpensively produced show. Let's assume the producers got (if they were very successful) $170,000 a show. That would mean that they made $40,000 net a week for 26 weeks. A nice business for you or for me, but distinctly limited. There is no overseas market, and most important of all, there is no residual used film business. The used film market is primarily in independents, and they've had the program already. They've used it and made whatever money they can.

Now let's say you do a program for the networks. Even if you make no profit on your network run, and you make only a couple of dollars in foreign sales, the end of the rainbow pot of gold is in U.S. syndication for those enormous numbers we've discussed. So the creative guys, the guys who really can put shows together, say, "Hey listen, what the hell am I doing making *Mary Hartman, Mary Hartman* when I could be making *Charlie's Angels*. You make *Charlie's Angels* and at the end of 5 years of production totalling 110 episodes you're going to make yourself $20, $30, $40 million. Other than a Norman Lear who is turned on by doing something like that and has the incredible ability to do 14 things at the same time, new programming for syndication is not really a marketplace.

QUESTION: You talked about a videotape show such as *Mary Hartman, Mary Hartman*. How about something like the Landsburg *In Search Of* series?

NORMAN HOROWITZ: That would be different. That would have at least a limited overseas market, and there would be money coming in. For an independent like Alan Landsburg, this is significant money. Part of the business is that you throw around numbers that for me, for anybody, would be wild, terrific. They may net $10,000 a program in the overseas distribution of *In Search Of*, if they do well, and that's a lot of money for Alan Landsburg, a lot for me, a lot for you, but again, for the type of numbers we've been talking about, not really a lot.

QUESTION: Do producers on the level of Norman Lear have their own syndication companies, or do they go to an independent syndication company?

NORMAN HOROWITZ: When the rights to *All in the Family* were extracted from Norman Lear by the network he was not able to do that. His company, Tandem Productions, actually entered into litigation to get back the distribution rights to *All in the Family* from CBS, which demanded them in exchange for putting the program on the air in the first place.

Since then he has, in my view intelligently, formed his own distribution company with the help of Jerry Perenchio, the brilliant business head of his company. They had two very good reasons. First, of course, they wanted to keep all the money for themselves rather than give a cut to an outside syndicator. But secondly, and most important of all, you can't trust anybody else with that level of asset. I wouldn't. Even if I knew nothing about the business, I would find out and then I would go into the business. And that's what they did.

Think for a minute about the figures for a show like *All in the Family*. There will be about 200 episodes altogether. Take $500,000 per negative as a reasonable expectation. That's $100 million. Let's say Lear is then caught in a 40 per cent distribution fee deal, which leaves him with a $60 million net. Take off distribution expenses, residuals, one thing and another from that, say another $10 million, that leaves $50 million, of which I guess Norman Lear gets about $30 million. Even with the deal that was forced on him by CBS, he is not doing badly. My company tried to buy Tandem a couple of years ago, and it would have been the best deal we ever made, but he didn't want to sell.

QUESTION: *Mary Hartman, Mary Hartman* was independently produced, so he handled his own syndication on that?

NORMAN HOROWITZ: No, not on *Mary Hartman*. With *Mary Hartman* he went around looking for a sucker. He had a $30,000 or $40,000 a week deficit, and he said, "Why the hell should I take the deficit? I'll look around for some dumb distributor who will do that." For a position on the up side he found a dumb distributor, who fortunately wasn't me. Then as soon as it succeeded, he took it away from that distributor, and either handled it himself or went to somebody else. Very, very smart people.

What's Hot?

SONNY FOX: Looking at what is on the air in the 1978-79 season that has not yet been sold into distribution, what do you really want to acquire? What show are the syndicators chomping at the bit to distribute?

NORMAN HOROWITZ: The hot item that's out there not committed is *Three's Company*, the dirty show that precedes *Soap* on ABC. *Soap* got all the heat about immorality from the press and the public, and *Three's Company* is the dirty show that's a tremendous hit and is not controlled by a major distribution company. The guy who controls it is Donald Taffner, an ex-agent for William Morris. He was in the distribution business representing Thames Television, a British broadcasting organization. They had a show called *Man About the House*, on which *Three's Company* is based. He now has a 35 or 40 per cent distribution fee on that show and he's not really in the distribution business. Guys are coming up and saying, "I'll buy you out for $2 million," or "I'll buy you out for $4 million," and those are nice numbers. I offered to buy him out too, but if I were Donald I would hold on to it. I would keep putting off making a deal, but then again, the dynamics of the business are such that suddenly, what is a 40 share show today goes against tomorrow's terrific show and dies. What happens to *Three's Company* if Suzanne Sommers runs away tomorrow, doesn't want to do the show, and they replace her with some new actress nobody can stand. Donald Taffner has turned down offers of maybe $4 million, and has no hope in the world of making anything like that if the show suddenly bombs.

I'll give you another example. Quinn Martin produces all kinds of detective shows, cops and robbers things like *Cannon, Barnaby Jones* and *Streets of San Francisco*. He has held on to the distribution rights to most of these shows. This should be worth zillions of dollars. Then along comes Family Hour, and all of a sudden you can't give these violent shows away. The Family Hour limits what you can program in that 7-8 o'clock time slot, and that's when the local stations are buying syndicated programming. Family

Hour cost Quinn Martin, I would guess, $50 million. It's a funny business—it's a gamble.

The dynamics that exist in the marketplace are such that someone like me, who won't play poker for more than 10 and 20 cents, finds himself in games where the stakes are enormous. I had a funny situation the other day in an Australian deal. We had made a deal with a station in Australia for a group of features for $1,855,000, a nice round number. They came back to me, even though we had already agreed on this deal, and said we accept the deal, but only at $1,800,000, knocking $55,000 off the agreed-on price. Now nobody in his right mind would blow a $1,800,000 deal over $55,000, but I said on the telephone to one of my associates who was handling the deal for us in Australia, "Tell them no." It was really funny. Here I was, with my company's money, saying no to $1.8 million as a matter of principle. It's like the way people shop for coffee for an extra hour to buy it for 7¢ less a pound. I'm very happy to say that they called back and confirmed the $1,855,000 price. That's part of the exercise of power in this business.

SONNY FOX: You get the sense, I hope, from what Norm Horowitz has told you, that although we all get hung up on the networks, and the dollars and cents there, there's another kind of hidden distribution, hidden to the extent that many of us are not aware of it, and those of us who are aware of it often are not aware of the details or the dimensions of it. You're talking about a syndication value of $400,000 per negative for a show for which the network paid perhaps $150,000 per episode. Now $150,000 seems like a lot of money for one half hour, but it's peanuts compared to what ultimately the residual value is in that negative, if the show is a hit, if it stays on the air for 5 years.

At the same time, we've heard a lot about the successes, but you should keep in mind all the shows that went on the air and didn't make it, and all the money that was put up by the distributors in advance in the hope that they might make it. That represents a lot of money and a lot of failure in there too. You always hear about *My Fair Lady* on Broadway, but you don't hear about that dreary parade of shows that never quite made it through the first Saturday night.

The Role
of the Agent

Rowland Perkins

In 1975, Rowland Perkins joined with four of his fellow agents to found Creative Artists Agency, serving as its first president and helping to build the company in a short time into one of the major talent agencies in the industry.

A native and long-time resident of Los Angeles, Perkins is a graduate of the University of California, where he majored in business administration (finance) and minored in theatre arts. After a stint in the Navy/Air Force and a period in law school, his professional career began in 1959 when he went to work as a trainee at the William Morris Agency. He served there as an agent, then was made head of the TV Talent Department, and at the time he left the agency, was Vice President in charge of the Creative Services Department on the West Coast.

I thought I'd cover the world of the agent from two areas: what an agent is and what he does. He's really the least understood person, I think, in the business today, maybe by design, maybe by accident.

The daily work of the agency and any given agent in particular, working for a full-service agency, covers a wide range of things. It covers handling individual clients, counseling them on their careers, and first and foremost, getting them employment. It means handling directors, writers, actors, producers, and, in my case, packaging, which I will get to a little later. You deal with studios, networks, handling negotiations on individual levels and on package levels. You meet with the business affairs people following your meeting with the sales presentation executives, who obviously have shown excellent taste in selecting your clients' projects to be made into shows. There's a meeting.with the clients' lawyers, there are meetings with their publicists and the business managers who handle all their financial affairs. You have to be with them socially evenings, just for a little hand-holding. Really being an agent, or a successful agent anyway, is not a job, it's a way of life. It consumes most of your days and weekends, but obviously it's satisfying or I wouldn't stay in it.

Maybe I can give you a better idea of the work by telling how I got into the agency business and what happened to me. In 1959 I graduated from UCLA with a degree in finance, a minor in theater arts, and a year of law school, with the idea in my head of being a producer. Fortunately, I knew a number of people in the industry and I talked to them about it. A lot of them told me, "Well, you don't just walk out and produce films." They also felt that I might get a good overview of the business by working with an agency, since I had the right temperament and personality and they thought I would do well. If I did, after a few years I would understand the business and then could go out and become a big-time producer. I met with four different agencies where people had set up appointments for me and, as luck would have it, I had offers from two of the major ones—MCA at the time, and William Morris. Knowing very little about either of them, I discussed with the same friends exactly what I should do. The consensus was that I should probably go with the William Morris office because they felt I would get along better there. I did that, and it turned out to be a wise decision, since 2 years later MCA was forced to divest itself of the agency business due to a Justice Department anti-trust ruling. MCA then was made up of Revue Productions, which was a television producing arm, Universal Pictures, Decca Records and a few other things, and the agency business, as big as it was, only brought in 17 per cent of the profit for the corporation. So it was an easy decision for them.

I started with William Morris in their mail room at the magnificent sum of $45 a week, with my parents asking themselves what kind of a loon they'd raised and put through college. But I stayed in the mail room about three months, delivering packages and things around town with the other mail boys, and really, you learn a lot doing that. I learned where the studios were, who was involved and I began to feel comfortable with the "town." I then was made head mail boy, or dispatcher, which meant that I didn't run around, but sent the other boys running around, except for emergencies. I did that for about another 3 months. While I was doing this, I was going to business school to learn shorthand, or speedwriting—I already could type—and then they assigned me as secretary to one of the agents.

I was assigned to a specific agent who, at the time, was in

charge of packaging for Four Star Television, then in its heyday. It had as many as 12 series on the air at one time, and they were probably the agency's biggest client. I worked with him for about a year and a half, really becoming an assistant during that time, and had another secretary working for him and myself. When I came back from a vacation I was told that he had another assistant, and I was going to be an agent functioning in the talent department. For one day I went out with another agent to a couple of studios, and from that point on I was on my own.

I then worked for about 3 years as a straight talent agent, selling basically actors and directors to the various studios, and then I was asked to become head of the talent department. For a period of about 5 years I ran the talent department, supervising the other agents and continuing to make deals for my individual clients, along with the others. I then was asked to reorganize the television packaging department. Why they picked me I don't know, but fortunately for me, they did. Here was the biggest, or second biggest, agency in the business, operating its packaging department on a hit or miss basis. I had a problem because I was running this whole packaging department by myself. I had to meet with the clients, pull the packages together creatively, and then I had to sell to all three networks. This caused some problems. For example, this was when Herb Schlosser was with NBC on the coast. He happened to know that my personal relationships were strongest at CBS and he'd say when I presented him with a package, "Oh, CBS didn't like it, huh, and you're coming around to shove it off on me?" We'd kid around, sort of on the square. I realized that it was a practical as well as a personal problem, trying to do it all alone. So I convinced the company to let me bring a couple of associates in from other departments of the company, and I slowly, with them, built the department up until it consisted of five pieces of manpower, including myself.

I then broke the networks down and gave each one sales responsibilities for a network, so that relationships could be established, and all of us then would meet with our writer clients or producer clients to try and develop shows.

After that period, which ended in 1975, I left with four of

my associates, and we set up our own agency, Creative Artists Agency, which has done, fortunately, very well in the first 3 years. We left with our briefcases in our hands and our hearts in our throats, and without any clients, but knowing that we were going to pick up some clients we had been close to through the years. We then moved into offices and in a period of about 2 months we acquired about 40 clients, most of them being clients that we were close to. Michael Ovets, one of my partners, was probably one of the best agents at that point in the daytime area. We handled several companies—Jack Barry–Dan Enright Productions, Bill Caruthers Company—and we were very fortunate that he sold three game shows, and got them off in three months. So we had a little bit of income coming in. Since then we have moved into large quarters in Century City, and we merged our company with Martin Baum's company. We now have six partners. Marty Baum had been president of ABC films and, most recently, before he went back into the agency business, had been producing independent films. Prior to that he had been head of the CMA motion-picture department, IFA and GAC, and had his own company in New York. He strengthened us quite a bit in the motion-picture area. We have since hired ten agents to work for us in various capacities. We have also added two business affairs lawyers who were with the Morris office as head of their legal department, and we have three accountants, so we are now a good, solid, medium-sized company, with a total of about 40 people, including ourselves. We're probably already the third largest agency in the business, in only about 3 years. Anyway, that is my personal career.

Now let's talk about packaging. Primarily this is my function within the company, along with two of my associates. I still make talent deals for clients I have handled for a long time, but primarily I'm in packaging. I will start with two situations of individual clients that I package from a talent point of view, and then I will talk about one specific show situation and carry it through from the beginning to the end, showing how we're involved.

Packaging William Conrad

One of the talent deals that I'm currently involved in is for William Conrad, the star of *Cannon*. Fred Silverman had found Bill, who had been an actor and had gone back to Warner Brothers where he was producing and directing. Bill ran a unit for Jack Warner and was directing also, but he had gotten out of acting. When Jack Warner sold Warner Brothers, Bill was out of a steady job. We talked·about it a little bit and he went back and did two television shows. He did *Name of the Game* and *High Chaparral*. We also agreed that if he was going to act, even though he hadn't acted in 12 years, he would only take leads. A show came along where I had an offer for him. It was one day's work on a Movie of the Week for CBS called *Brotherhood of the Bell*, and Bill didn't want to do it. I really felt he should because it consisted of a 14-page scene with Glenn Ford. It really was a terrific role. We went round and round about it and he finally agreed—he asked me to ask them for a lot of money—I forget exactly, but something like $10,000 for the day. He figured he couldn't argue me out of it so he just made a demand of a high price. Well, they came up with the money!

About 3 months later I was having lunch with Quinn Martin, because I was also handling him, and he said to me, "Rowland, I want to level with you. I want Bill Conrad for a show. I have a pilot with CBS and they told me if I get Bill Conrad they'll go to film on it." It all came off of that *Brotherhood of the Bell* that CBS had seen. The show was *Cannon*. Quinn said to me he would make any deal that the network could swallow, so the deal was made for Bill to go into *Cannon*. He was not the conventional leading man. He was not Robert Wagner—he was not that, but it just worked. The public identified with him.

Cannon was a very successful show, as you know. It ran 5 years and made Bill somewhat of a television star. When the show was cancelled, Fred Silverman, who felt Bill was his discovery, had moved over to ABC. Fred called me and said, "I would really like to make some kind of a deal with Bill." He basically wanted to show the world he made Bill a star once and he could do it again with the same guy.

When we made the deal Bill said, "I'm going to do this myself. I'm going to go into business for myself. I'm going to own the show, I'm going to have the profit participation, and I'm going to have the control."

Let me take a moment to tell you about Bill. Fortunately, he became very friendly with Fred during his time at CBS, so much so that they even took one vacation in Hawaii together for several weeks, with Fred and Cathy, his wife. To show what kind of a playboy Fred is, Bill said that Fred spent the entire 2 weeks on the phone to the coast and to New York, or watching television. He didn't go to the beach, he didn't even go out except to dinner. Absolutely incredible.

But anyway, back to this. Bill also, because of his draw at CBS, does the Thanksgiving Day Parades and Cotton Bowl Parades for CBS, even though he's got an ABC deal now; they still want him to do it. Bill was back East doing the Thanksgiving Day Parade and Fred asked him up to have dinner. Now basically I had made the deal for Bill's company under which ABC was guaranteeing him a movie for television, a couple of pilots for his company, and whatever else was involved. The deal was closed. Bill and Fred were having dinner and sitting around, having a few drinks, and when they started talking, they created a television show. Fred told Bill, "You go back home and call Steve Gentry and Brandon Stoddard, who are running the coast here for ABC—you tell them this idea and they're going to love it." So we immediately go in, and Bill told them the idea, which they loved, and we put it into development.

This isn't really what I normally do as agent, but there's a little thing here which is an interesting insight into a network. We now had to get production people, because Bill was not about to go on the line to produce. One thing Bill's learned is what he enjoys doing, which is making a lot of money and working as little as possible, so we tried to put together a deal with Jerry Thorpe and Bill Blinn, with whom ABC had deals. We couldn't really work that out. I then went and made a deal with Everett Chambers, who was a *Columbo* producer and who was also under contract to ABC, to come on the show. We arranged a meeting for Bill Conrad and myself with Brandon Stoddard, vice president in charge of drama, and Steve Gentry, and talked out this whole show.

Cliff Allsberg, who was in charge of drama development at ABC, was not at the meeting. OK, once we've had this meeting, we go away and hire a writer, Dick Jessup. We bring him in, he does an elaborate 70-page treatment, and we deliver the script to Cliff Allsberg, who loves it and tells Jessup to go back and write the script. Jessup, who did *The Cincinnati Kid* and other films, is a good writer. He now goes away and writes a screenplay. We then deliver this to Steve Gentry and Brandon Stoddard.

Next, I get a horrified phone call from them. "This isn't what we talked about!" I say, "Wait a minute. We sent you a 70-page treatment a month ago. Now we did nothing but add the dialogue to the script. There's nothing different!"

Finally they explained that neither of them had read it, and Cliff Allsberg, who had not been at the initial meeting, had assumed that the 70 pages followed what we had talked about at that meeting. So, cut. We lose a pilot because we can't get into production in time. But we still have a deal. So now we start all over in midstream. We have to go back, have meetings, insist that Stoddard and Gentry are there with Allsberg, spoonfeed pages to them. The script was just delivered this morning—I don't know what's going to happen.

This is one of the things an agent has to put up with in dealing with a network. And the irony of it is that this is costing ABC about $120,000 in excess payments that they have to make for holding up the production company for 5 to 6 weeks while the new script was being written. It's really crazy. Anyway, that's the kind of deal you make for an actor with his own production company.

Packaging Chad Everett

A different kind of deal that we are working on right now is at CBS with Chad Everett, who was the star of *Medical Center* for 7 years, and who didn't want to do another series. CBS, about 2 years ago, paid him a lot of money just to do nothing—if he did anything he had to do it for them. If he changed his mind and wanted to do anything on TV, he had to do it with CBS, but otherwise he was free to do whatever he wanted and they paid him anyway. Finally, about the time the deal was over, Chad got a little bored, and I guess his picture career wasn't going the way he wanted. (All these television actors feel when they come off a series they want to be in the movie business.) Well, we went to all three networks and said, "Chad has now decided that he wants to do a television show." CBS guaranteed him a series, in essence—they wouldn't guarantee production, but they guaranteed that Chad Everett would be paid a very substantial five-figure sum per episode, for a full order of shows, as his personal salary. And they will develop with his company, with proper people, shows to fill that commitment. They have also guaranteed him movies for television. So now it is incumbent upon us as agents to follow up once these deals are made with Bill Conrad and Chad Everett.

It's one thing to make deals, it's wonderful, but you have to then follow through and make these deals meaningful. You've got to get the right people. You search your own client list first, though many times you don't have the right ones, and then you start dealing with other agents to find creative people—writers or producers. In Chad's case we were also able to put a client in there, Frank Glicksman, who had been executive producer on *Medical Center*. He has a strong relationship with Chad, and we have them working together now developing a show, with Charles Larsen, the writer who created *Twelve O'Clock High* and other shows.

But meanwhile we still have to find other projects for Chad. You see, the problem is you make these deals for your clients and they're not really going to fill them themselves. They're glad to have the deal, but the agent then has to become the producer. The agent has this problem—and these are only two of ten deals I may have going on at the same time. I have to be juggling these things back and forth to give my clients the proper time and also service these shows.

We had the same thing at CBS with Sally Struthers when

she came off *All in the Family*. CBS didn't want to lose her, and they've guaranteed her a series, which we're trying to put together. CBS gave us the same deal with Stockard Channing, and I finally found a movie project, after many meetings. In selling the concept of that I had to call other agents to find writers. You make up lists yourself, creatively picking certain writers you want, then you find out they're not available.

Szysznyk

I spend half my day, in essence, executive producing, I guess. Let me take a specific show to give you an idea. This is a show that was on CBS in the summer of 1977, was then on again in mid-season, and subsequently was cancelled. It's a half-hour situation comedy starring Ned Beatty entitled *Szysznyk*. It really started one afternoon when I was having lunch with one of my clients, Jim Mulligan, a writer, and a friend of his named Ron Landry, who was half of a radio team on the West Coast called Hudson and Landry. They had an idea for a series about an ex-Marine who, after 25 years, has gotten out of the Marines, is now in his forties and his first civilian job is running a recreation department in a school. He had never lived as an adult in civilian life and he's trying to cope with that. It was set in a place called Turkey Thicket in Washington, D.C., a real place where Jim and Ron had grown up as kids. We talked out the concept a little bit at lunch, felt we had a good idea, and I gave them some thoughts about what the problems would be. About two weeks later they sent me a fully developed presentation, with the characters delineated, with the format, the sets, everything done.

When I read it I felt we should do some more peopling of it. It was a really good project for Shecky Greene, who is my client and who the networks had been trying to get into a television series for some time. He's one of those fellows who's been fortunate financially—he probably makes $2 million a year in Las Vegas—and so money is not any criterion. I think he's a very talented man and one of the funnier comedians, and also a good actor. He was once on television in a show called *Combat*—he had the third or fourth lead in it. Like all comics, he really wants to be an actor, yet on the other hand he doesn't want to, not because he's scared of acting, but he's afraid he's going to damage what he has, and I think correctly so.

Anyway, I called Shecky and told him the idea. In this particular script the lead is married to a Japanese girl, and as it happens Shecky is married to a sweet young lady who is Hawaiian-Chinese. So he said, "OK, I'm interested. I want to meet with the producers." Well, I didn't have any producers, so I talked with Mulligan and Landry and suggested that we bring in Rich Eustis and Al Rogers as producers. They happened to be clients, and at the time were working with Jerry Weintraub (who is a partner of John Denver and manages Frank Sinatra, Neil Diamond and others) in a show for ABC called *Father, Oh Father*, a pilot. They had worked with Shecky when they produced a Glen Campbell show, and they also knew Jim very well, because he had been head writer for them when they produced the Rich Little variety show.

They liked the idea a lot. They said OK, we're in. Four's Company Productions, a partnership between Eustis, Rogers and Weintraub, became the producing entity and Jerry agreed to take on the financing position in the project. We were now ready to go to the network, and so it was a matter of selecting a network. I picked CBS because I knew that they had been after me for Shecky for some time, so I knew there was an ongoing interest there. Also I knew that CBS liked Rich and Al, because they had been trying to make some other deals with them. We went in and met with Andy Siegel, who was head of comedy development at CBS, made our pitch and left the presentation. About three days later I got a call from Andy saying, "I'm interested, but I want another meeting, and I'd like Shecky involved." I called Shecky in Las Vegas, he came down with his entourage and we set up the meeting.

Let me say parenthetically here that this was the biggest meeting of its kind I have ever been in, and I'd never like to do it again because at large meetings it's almost impossible to sell. We had about 15 people at this meeting. It was so big we had to use the CBS conference room, Shecky thought he was in Las Vegas and started performing, but it was terrific, and he sure didn't want to pass up a house like that. He had everybody on the floor.

Anyway, we left the meeting with a pilot script commitment and a deal for Shecky Greene, which, by the way, he never signed. We went away, they wrote the script, came back, turned the script in to the network, put the script into rewrite and the network made some changes which they felt for their purposes were correct. We then delivered it, and they loved it and said, "You've got a pilot commitment." Joy was rampant until I gave the script to Shecky. He didn't like it, and I'm not sure to this day whether he really didn't like it. He had some valid criticisms, but I felt maybe he was getting cold feet about getting into television. Like a lot of comics they want it, but then they're more afraid of failure, and they're sensitive in their little clique. He was afraid of Don Rickles jumping all over him, afraid of Buddy Hackett and guys he's close with—it's more real than you might imagine.

In any event, I went back to CBS to break the news, wondering how in the heck I was going to do this. Fortunately I found that they were so happy with the show itself conceptually, that if I could keep it together, they would go with somebody else—if I could find somebody. So we looked around, went to a lot of people and finally were able, with a lot of difficulty—I won't bore you with details—to get Ned Beatty. They said, "Terrific, we'll go ahead."

We shot the pilot and it was quite good. At that point, I had to take the pilot, along with a couple of others to New York, which, at that time of year—pilot season—was an interesting experience in itself. I'm really not sure if there's any value in these trips, to be perfectly honest with you, except that everybody does it, so to stay competitive, or at least feel you're being competitive, and make your clients think you're being competitive on their behalf, you really have to go into New York for those lobbying sessions. In fact, at that time of year the bar at the Sherry Netherland looks like the Polo Lounge. You don't see a New Yorker—it's all West Coast guys. You sit there listening to them lying about how well they're doing, how well they're being received. You know, you can play games with some of the more neurotic types. You start making up rumors—you say this and that, this looks hot, or you let them hear that somebody just killed their children.

SONNY FOX: The problem with that, though, is you hear the same rumor coming back, and then you believe it.

The question of what you believe or don't believe in this business, who you listen to, is crucial to the success of a good agent. When I listen to an idea from you, Rowland, at least I know that I'm not listening to something that is coming off the street or through the transom, which makes no sense at all. It's gone through your filtration, your agency and your people. It has the ring of reality. Because of that, agents do have access to network executives, but only insofar as they've been able to prove themselves. There are agents out there I wouldn't spend the time of day with, because I really don't think they know what they're doing. They'll grasp at anything that walks by and they'll run in with it. Maybe something will click. They're as desperate in their way as any of the people out there are to sell whatever they're creating, so they really don't present much of a filter for us. So there are agents, and there are agents.

"Ten Percenters"

Now, one thing you've eliminated in your entire discourse is dollars, and we don't want to think that dollars are dirty around here. We're all interested in the dollar part of it.

ROWLAND PERKINS: I'm in it for art.

SONNY FOX: That's because you're already driving a Rolls-Royce, or a Jaguar, whatever that thing is. For those of us who are still in our American cars, why don't you talk about commissions. On a straight talent deal, when you represent talent in a movie or something like that, do you work on a straight 10 per cent?

ROWLAND PERKINS: Commissions really go into two areas. Somehow that phase of it I take for granted—people have run around having called us "10 percenters" for years. When you represent talent, at least in California, and I assume in New York, you work under the Labor Commission. Technically we operate under the same laws, same charter, same licenses as the agent you use to hire a maid, gardener, or whoever—an employment agency, because that, in essence, is what we are. So, by law, we are limited to charging a 10 per cent commission, which we often do, for any personal services. However, packaging is a different thing. Technically, I guess, there are no regulations on that. The maximum commission in the packaging area is coming down. We brought our commission down just for a sheer competitive reason. Also, with the costs going so high, the 10 per cent commission on a package really becomes a burden to the producers. The William Morris office will not reduce its commissions, one of the reasons I left. I had a very large battle over my insistence that it was imperative to be fair to clients and cut down our package commission and restructure it. I was informed that would happen "over their dead bodies." Since I didn't want to go up for murder, I decided that I was either right or I was wrong, and if I was right I would go out and prove it for myself. I obviously had some support because a number of associates went with me. Anyway, William Morris now takes a 5 per cent current and a 5 per cent deferred commission, which comes out of first profits. I thought that was onerous. As a good example of how much money can be involved, let's take another example of a package I put together back in my William Morris days, which was *Mod Squad*, a show that ran on ABC for a number of years. I don't remember what the license fees were on an hour in those days, but the Morris office took their 5 per cent deferred and they got this 5 per cent in first position out of profits. The way that worked out, the William Morris office took $1 million commission out of the profits before the producer saw a nickel, and I just felt that was outrageous. Our position today at CAA is that on the packages we take 6 per cent reducible to 3 per cent which is deferred, and we take the 3 per cent deferred out of 50 per cent of the profits. This means that of every dollar that comes in, the producer, or profit participant, takes half of it, and we take the other half until our 3 per cent is caught up, and from then on we take 10 per cent of additional gross sales. That's how we've structured our fees, although we've also been known to defer taking our share if the producer really needs it.

For argument's sake, let's say a license fee nowadays is in the area of $275,000 for an hour show, and depending on the elements you can even push it close to $300,000, but that's pretty high. So, as you can see, when you're talking $275,000, even at 3 per cent, you've got $8,250, a pretty good fee, once the show's going. You do have to police the show once it's going, but the hard work is getting it sold, getting it put together and then it's something your service people, with you just overseeing it a little bit, can take care of. Your business affairs people and accountants work with the show people.

So after a show is off and running, if it's a hit, your work's minimal, and in the long run you can make a large amount of money. To give you an extreme example, we've sold a show for a million dollar advance, made from a novel called *Shogun* by James Clavell, whom we represent. If this thing plays out, we will get about $900,000 in commission. It may not go. Nobody could sell that book. Jimmy would give us twice what we're getting for selling it—that book's been around. It was never even approached for movies. I don't know why.

SONNY FOX: Let's clarify this for a moment, because you've used two numbers—a million advance and $900,000 for your agency.

ROWLAND PERKINS: The million's an advance. It is what he got just for the rights to the book, and in addition he gets a piece of the profits. But we also got out of NBC the highest price they've ever paid for a show. We're getting a million dollars an hour, and will do about 15 hours. However, even at a million dollars an hour the producers are concerned because we will probably have to shoot in Japan (it's set in 16th century Japan) which means expensive costumes. We're obviously going to have to get a major star to play Blackthorne, the lead, to justify the investment. Norman Rosemont, who's a client of ours, is going to produce it with

Paramount. Paramount is taking the deficit responsibility, and will be backstopping the show. Norman will produce it through his company as a partner, and the whole project is going to be a 2-3 year project for everybody. But the network did pay a million dollars to Jimmy for the rights, and he was paid up front—he's got it now. He took it in two installments, spread it over two years.

SONNY FOX: That's nice. Cuts it down to only $500,000 a year. Let me ask you a question. Let's say I'm the producer, or the author, or somebody of that sort, and you're taking a piece of my action. Now I hear that you're putting Norman Rosemont in, who's your client, and maybe later you come in with a terrific talent who happens also to be your client. How do I ignore the feeling that maybe you're using this to promote CAA clients, rather than to do the best thing you can for the show?

ROWLAND PERKINS: Because it doesn't matter. When we put together a package we do not take a commission from the clients at all. You can't by law—not that we would anyway. For instance, we take a commission from Jimmy Clavell on his million dollars now. If they go into production we refund his commission. It may seem like a lot of money, but a tremendous number of man hours and expense, a lot of hard dollar expense, went into that. If we had not sold it (and there are many things you don't sell) we would have made nothing, so it sort of amortizes itself out.

Contracts and Releases

QUESTION: Do you usually have a contract with your client? How does a non-productive agreement end if the client feels he is not being served?

ROWLAND PERKINS: Yes, we usually have a contract. The only assets you have as an agent are your contracts with your clients. There are a few exceptions. Like, with Sidney Poitier we don't, because Martin Baum handled him for years before he came to us, and he's the kind of guy you never worry about. But yes, mostly. Mainly it's in the client's best interest, because you have to register the papers with the various Guilds and they have to be signed. The stack of papers you sign with an agency is awesome, let me tell you. For a period I was having to countersign all of them, and it took all day long. You have to sign Screen Actors Guild papers with an actor, along with AFTRA papers and AGVA papers, general service papers, material papers — that's really a wad. Then they're registered with the Screen Actors Guild. A lot of times producers call the Guild, and if there are no papers on file they don't even know where to find the client. There's a service called "Agency" at the Guild, which I use frequently. If I have to find an actor, and I don't know who represents him, I just call there. Same for the Writers Guild as for the Screen Actors Guild. With the Writers Guild—and I think with the Directors Guild also—you can call up and get credits on anybody, if for any reason you need to know them. On a personal level, let's say somebody calls and asks if you want to represent an actor. Let's say you're not familiar with him. You call the Guild and by the time he comes in you can get the file and sound as though you really know everything about him.

Terminating authorizations depends on the nature of the termination. Frequently it will be a mutual agreement. If the client is unhappy because nothing's happening for him, you in turn are unhappy because obviously nothing's happening for you. In certain cases we've had discussions that run this way, "We shouldn't have signed him, we made a mistake, it's just not going to work, we don't have the time to give him, it's just not fair, but who's going to tell him? *You* call him." "No, *you* call him." Then, like a blessing, a call comes in from the client. "Can I come in and see you?" He's more nervous than you about it—"Gee, I hate to do this."

In some cases we have called the client in and he said, "Look, I know it's my timing, and everything else, but I can't change myself. I can only change my agent," and you say, "All right."

There are different kinds of releases. There's one where he comes in and says, "I want a release," and we give him back his papers along with a letter saying that our contract is null and void. There's one where the guy comes in and we say, "No." A lot of times it's arbitrary. "We've done a good job here, and you're doing this and this." Or maybe he wants a release because he'd been to a party and met some other agent, who—as we like to say—has done a good "shuck and jive" on him, and he thinks it sounds wonderful over there and he wants to get out, and we say "No." So, he can then discharge us, which means he sends us a letter notifying us that we're no longer to be his agents or hold ourselves out as his agents, and we answer that back with a letter saying that's fine, but we're holding you to the terms of your contract and we're entitled to commission on any deal you make until such time as your papers run out.

Even if clients leave amicably, we continue to collect on anything they have done while they were with us, including residuals.

QUESTION: When you say that you're going to hold on until the end of the contract period, characteristically what is that—one year?

ROWLAND PERKINS: Well, it depends. Take an actor, for instance. First time you sign an actor, by regulation, you can sign him for a one-year contract. After that, you can sign him for up to 3 years, on one paper. It used to be you could sign directors for 7 years, which is incredible. For writers, it's 2 1/2 years; 2 years the first time. As for directors, you can sign them for 3 years right away. I don't think you can sign anyone for over 3 years now.

Agents and Managers

QUESTION: What's the difference between an agent and a manager?

ROWLAND PERKINS: Basically managers deal with talent, mainly actors or actresses. The difference is hard to state. I don't want to say that an agent does not get involved in career discussions, because we do. But basically, an agent's prime responsibility is getting employment. By law, a personal manager is not allowed to get employment, though some of them do in a sense, but they can't close deals, technically. Some lawyers are starting to get into both areas—agenting and personal management. Even some publicists have gotten into management, like Jay Bernstein with Farrah Fawcett and a couple of people for whom he's doing publicity and also managing them.

An agent basically gets employment. That's what we're there for, and we do have some management functions. I would say far more of our clients don't have managers than do. Management is not regulated, and if you want to get 90 per cent you can. I know Colonel Parker, when he handled Presley, took 25 per cent. So he would take 25 per cent, an agent 10 per cent, a business manager 5 per cent. Presley probably gave up 40 to 50 per cent of his income just to his service people.

Who Will the Agent Represent?

SONNY FOX: What about Farrah Fawcett-Majors, who you mentioned? I think it might be a very healthy thing for you to make a public confessional here about some of the biggies that got away.

ROWLAND PERKINS: All right, I'll do that. I was called as an expert witness in the Farrah Fawcett case, because I really sort of deplored what she did. If you haven't read about it in the paper, Farrah Fawcett decided at the end of the first year of *Charlie's Angels*, when she had gotten hot and was in demand, that since she didn't have a contract, she didn't have to stay. Technically, I guess, she had never signed a contract, but she had worked for a year and taken the money. I think it was wrong because she was just taking advantage of the situation. Anyway they'd asked me to come in and testify. I was going to do it until my partners talked me out of it. I guess I understand their feeling. Their basic attitude is practical rather than idealistic, and I understand it when they say, "One, we'll never have a chance to represent her, even if we would want to, and I don't necessarily want to. Also, her attorney's a close friend and he's got a lot of big clients—and you'll probably never get a chance at them." So I guess that's more pragmatic. Anyway, it's true.

Judgement comes into it a lot for an agent, particularly when it comes to talent of actors—which you represent, which you don't represent. A few of the people the last 3 years that we have turned down for representation were John Travolta, Farrah Fawcett, Jaclyn Smith, and a couple of others. There were valid reasons at the time, and if the exact set of circumstances happened again, we would probably do it again. When Travolta's manager offered us his representation, John was in the first year of *Welcome Back, Kotter*, and we had just set up our company and needed cash flow people. He was under contract that could have gone 5 years and we'd never see a dime, and we would have had to devote a lot of time to servicing him and everything else. Nobody questioned that he was an interesting, talented guy.

Within our company, we vote on who we take on and who we don't. Any one of us can come in and plead our case in front of the others as to why we should take on someone. More often than not everybody says fine, if someone's really impassioned about it. But frequently, we debate it and literally vote on it and either the yesses have it or the nos have it. We've tried to run it as democratically as possible. We also have in the company what the partners call the "ultimate veto." We each have one a year, if we want to use it. No matter what it is. If one partner says, "I'm exercising my ultimate veto," we all go along with it. But then, none of us has done it in 3 years. But you save it—you never know when you're going to need it later.

SONNY FOX: Let's get to some tough questions. What does it take to be taken on by CAA now? You've got a very successful agency now, you're doing very well and somebody walks in the door and says, "I want you to represent me." What are you looking for?

ROWLAND PERKINS: We've had to become selective. First of all, I think you have to look at an agent or agency and say, "What is it?" It's pure personal service, and what you have basically is your reputation, your credibility with the buyers as to whether they accept what you offer. You have your time, and you have whatever expertise you picked up over however long you've been an agent. And you can't really squander that, so you look at several areas. One, what is the potential of the property, whether it's an actor or director, what is your evaluation of the potential? I think this is the first thing. Will there be a career? Will there be an economic return? Basically, can we do a job for him?

A lot of talent we've turned down because we don't function in their areas. We have taken the agency totally out of the episodic television business, mostly for economic reasons. We may be doing ourselves a disservice, but we don't want to get so big that we lose what we have tried to set up—a personal business with good relationships. William Morris has something like 60 or 70 agents who don't even know half the clients there. So we don't want to get that big. We only have so much time and we want to devote that time to setting certain priorities. So, if an actor is basically going to work in the episodic business, even if we think he's got a future, we have to say no at this time, because we're not going to get any of that work, and we know it. It's not fair to the actor to think that we're going to.

One luxury we allow ourselves, particularly with established people is, if we don't like them, we won't take them on. We didn't have that luxury when we were with the William Morris office. Usually somebody would be signed up in New York, and bingo he'd get on a bus, and soon be sitting here on the coast. And he'd be somebody we wouldn't have wanted to sign. Having a choice is a little luxury, and I'm happy that we do have sufficient people earning enough income for us and for themselves so we can turn down some people.

SONNY FOX: What about packages? A guy walks in off the street, and says, "I've got this terrific thing, it's 70 pages." Would you read it?

ROWLAND PERKINS: 70 pages, no—61 is our limit.

No, not off the street. You just have to stop after a while. You can't read everything. First of all you might get into trouble, get into lawsuits. People submit scripts—and nothing's new under the sun—it's always derivative, or a variation on something else. You read something unsolicited, and turn it down, and then, say, a client who's very established comes in with something similar and you package it. You either have forgotten what you saw, or maybe you remember, but this unknown guy hears about it, and the next thing you know he's slapping you with a lawsuit that you stole his idea. You have to get releases signed before you start. Anyway, I've basically given up taking in stuff off the street. It usually has to come from a recommendation of somebody.

How Much Service?

SONNY FOX: Another tough question. You have guys who are spinning off a lot of money for you. Shows, and also talent—people like Rod Steiger, that sort—and you have people who aren't spinning off that much money. Are the services provided to these people roughly proportionate to their value for the company monetarily?

ROWLAND PERKINS: Not the services, but maybe the amount of time you devote to it. The quality of the service doesn't vary, but the time, sure. It's only natural that the demands are greater with a working star. It's not that you say, "I'm going to spend more time on Rod Steiger, or I'll spend more time on Richard Harris, and less time on Karen Valentine or Barbara Feldon"—it's just that either the number of things you're doing take more time in negotiation, or you have to go to Europe because he's shooting over there on a picture and you have to see him. It's nothing calculated and you don't put your clients into categories.

QUESTION: Take one of your clients who is not currently a hot property, but has been hot in the past, and he knows you're not going to devote a lot of time to him. What's your attitude if the client goes out and promotes himself, acts as his own catalyst, gets employment and says this to you, "I've got this person who's interested in doing this with me." You're asked to come in from that point on.

ROWLAND PERKINS: That happens. How a client helps himself is an important factor. One thing that anybody in this business had better realize is that an agent can only do so much. If an actor is going to sit home and wait for the telephone to ring, he's going to be in a lot of trouble. You can't do that. You have to do certain things for yourself. A writer, at least, can sit down and write. You know Paddy Chayefsky can give you a piece of garbage just as easily as somebody else, but you're going to be afraid to admit to yourself that it's garbage if his name's on it. But I can't hit every time. At least with something on a piece of paper I just try to read it for whatever it is, and you either like it or not. And then maybe I'll find out it's Paddy Chayefsky or just some guy.

I'll give you an interesting example. I had a presentation given to me by a credible source. He didn't tell me who wrote it, but I read it and it was really good. When I told this guy my reaction he said "I thought you'd like it—it was written by my handyman." The handyman was writing on the side, and he was really good. He didn't dare tell me up front because I would have laughed. I showed it to a couple of my partners and said: "Here, read this, I think we can do something with it." One of them asked me who did it and I said John Kulhanek. That was the guy's name—I figured nobody would know him, but one of my partners broke out laughing and screamed, "That's Joe Blow's handyman!" Once in a blue moon that would happen!

Cash Flow

QUESTION: How does the flow of money coming in and going out, the job of controlling it, compare with, say, a production company? Are there special problems for an agency as against production?

ROWLAND PERKINS: It's a reasonably simple structure. Yes, there's a heavy cash flow, but we keep reasonably simple records. An agent, when he makes a deal, puts through a booking slip through his secretary, which then goes down and sits in our accounting office. Tickler files are set up for following up collection of those things, and each week the partners are handed an itemized income statement for basics, so that we know the cash flow for the week, and more importantly so we can follow the deals and collect not only for ourselves, but for the client. It used to be very seasonal, but it's leveled out. I think television with all its mid-season, and mid-mid-season and mid-mid-mid-season starting is becoming more all year round, and the movies of the week are constantly in production. We used to hit a real dry spell from about February to mid-May or June, because actors weren't working and television was shut down. But now it's pretty much year round.

QUESTION: You mentioned earlier that you'd rather be an agent than a producer. Just like you have to go to a gas station to get gas for your car, you have to go through an agent to get work in the entertainment business. That's a very lucrative business, isn't it?

ROWLAND PERKINS: Yes, it's a lucrative business, if you've got the right clients. It's a tough business, believe me, and the toughest part of it is getting to the point where you can attract the clients that are lucrative.

Getting Started

SONNY FOX: If somebody's coming out of school and wanted to get into the agency business, what would be the best way to do it?

ROWLAND PERKINS: You've got to work through one of the big agencies and go through a training program. They still give training programs.

QUESTION: I had a meeting with Dick Clark Productions, and they said "You should be an agent," so I called up William Morris, and they were really excited and said "It's $125 a week." I drive a Mercedes, even though I haven't really made it yet, have a big house—and for me I thought it was a wonderful idea, except what do I do with the $125?

ROWLAND PERKINS: It's one business you can't set up like other businesses—it takes contacts and it takes time to build them. You have to start at the bottom.

SONNY FOX: What you need is an agent to get that job. Wouldn't that be a funny gag? Call up William Morris and say you want the mailroom job for $125, please call my agent.

ROWLAND PERKINS: I can't think of any agents who have really made it who didn't start out with the proper background, so you meet the people in the business.

SONNY FOX: It's like being a page at NBC. Alan Landsburg told you he came out of there and so did a lot of other people. Sometimes you really have to pay your dues, as he said, and it's the only way to do it.

ROWLAND PERKINS: I really started late—I was 25 when I started, and they almost didn't take me then. But I'd been through school, had to go into the service, came out, came back and finished my education, went to law school. I started at $45 a week, big stuff, and I don't know how else to do it, frankly. We have a couple of young men in the mailroom at CAA now. They've just come out of college, and one of them still lives at home. The other kid happens to come from a fairly well-off family, so his father's subsidizing him for a couple of years. He's very bright and I think he's going to do well.

QUESTION: What about the clients who you never find work for? Do you still get paid some sort of annual fee?

ROWLAND PERKINS: You get paid for performance—if the client makes nothing the agent gets nothing. That's the case with a lot of the service people in show business. Publicists are on retainers, and they're usually the first to go when a client gets in trouble. Most everybody else works for a percentage. Agents get their percentage, business managers are usually on a percentage as well.

QUESTION: You were talking about schooling before. Do you see yourself going into academia like David Geffen, the record executive who also started out in the mailroom at William Morris?

ROWLAND PERKINS: I'd enjoy it, and I guess at some point I might take a sabbatical and do it for a while. It would be fun for a year or so, but I don't see it as my ultimate goal.

That reminds me of a great little story. I was at a big meeting at the William Morris office and they brought in Joe Papp, who produces a lot of shows on and off Broadway and did "Shakespeare in the Park." They were trying to sign him, and they were really out to get him. They had a big meeting, with Abe Lasfogel, Chairman of the Board, and about seven or eight agents. One agent, who is now president of the company, was in there and was really trying to impress Papp. He was going on and on, and Papp was sitting there, and I could see he's not a guy who's easily impressed. Finally this agent can't stand it, he's just got to say something, so he says, "Tell me, what do you want? What do you really expect? What's your ultimate goal?" The implication being, clearly, ask me and it's yours. So Papp just sits there and looks at him and he says, "Death." So Abe Lasfogel jumps in with, "Joe, how about another drink?"

SONNY FOX: Joe Papp was my floor manager when I did *$64,000 Challenge*. He told me then about this cockamamie dream he had about free "Shakespeare in the Park," and I just sort of humored him.

I hope you have all seen another aspect of our business through Rowland Perkins' discussion, and found out that not all agents are really vipers. Some, in fact, are really creative and very intelligent people.

Independents and O and O's: Image Problems

Lawrence Fraiberg

Dennis Swanson

Lawrence Fraiberg has been active in the television industry from its earliest days. A graduate of the University of California, he joined KPIX-TV in San Francisco in 1949, where he was a salesman, producer, writer and occasional announcer. He stayed at the station for 10 years, becoming General Sales Manager in 1958, and in 1959 left to join what is now Metromedia to assist in the acquisition of additional television stations.

In 1963 he was made Vice President and General Manager of WTTG in Washington, D.C., and in 1965 took over the same position at Metromedia's flagship station, WNEW-TV in New York. He left in 1969 to form his own motion picture and television production company, returning to manage WNEW again in 1971. In 1977 Fraiberg was named President of Metromedia Television, the position he held when he made the remarks which follow. Since that time he has left his position at Metromedia to pursue his own production projects.

Dennis Swanson has been News Director of KABC-TV, the ABC network-owned station in Los Angeles, since May, 1977. He joined KABC early in 1976 as Executive Producer for News, leaving Television News Inc., where he had been since 1973, both in Chicago and New York, as News Director.

A native Californian, Swanson received his B.S. degree in journalism from the University of Illinois in 1961, and his Masters degree in communications and political science in 1964. He began his broadcasting career as a reporter with WMT radio and television in Cedar Rapids, Iowa in 1964, and two years later moved to WGN in Chicago as Assignment Editor. In 1968 he joined NBC News in Chicago as a sportscaster.

LAWRENCE FRAIBERG: First let me establish what Metromedia is. It is a corporation that, among other holdings, has radio and television broadcast divisions. I'm president of the television division. We presently have seven television stations—five VHF stations, in New York, Los Angeles, Kansas City, Minneapolis and Washington, and two UHF stations, in Houston and Cincinnati. One of them, in Kansas City, is network affiliated. My role is that of administrator, setting policy and making sure that we keep our licenses. Up until about a year ago I was general manager of Channel 5 in New York, which is an independent station.

Perhaps at some point in time in the past, if given their druthers, the independent stations would have preferred not to be independent. They might have preferred to be network affiliates, since in the day-to-day run of things, it is easier to have a network provide you with anywhere from 65-80 per cent of your programming, leaving you to worry about only 20-35 per cent of it, than it is to program from sign-on to sign-off. It's also considerably more expensive to program seven full days every week, and, in the beginning, it was enormously difficult to find programming.

A major problem for the independents from the sales point of view is the fact that the advertiser has perceived the independent station as a second-class citizen, and in many ways still does today. There has been a concerted effort by the independents, certainly on my part when I ran a station and now as president of a group, to find a way to alter that point of view. One of the answers to improving the independents' status recently has been the establishment of viable news departments. About 10 years ago, Channel 5 in New York City started a 10 o'clock news, which was widely seen as not having a chance of really making it. Contrary to a lot of the speculation, it not only made it, but made it very well, and other independent stations in the country soon recognized how important news is as a window to the station—as a primary part of your image. Doing the news is expensive. It gets more expensive, because of the technology and other changes, almost monthly. But as the cost of other aspects of programming increases astronomically, news is proving to be not so expensive in comparison.

DENNIS SWANSON: As Larry indicates, news can be an image builder, and in turn can bring revenue to the station. I work as news director at KABC in Los Angeles, which is a network owned and operated station. The O and O's are a separate division of the ABC Corporation, and I might say that we are the most profitable division. There are five VHF stations altogether, in New York, Chicago, San Francisco, Detroit and Los Angeles. KABC-TV is the most profitable station in the most profitable division of ABC, and that is largely a result of the effort we've made in news. It was not so long ago that KABC-TV news was in third place in Los Angeles, but now the worm has turned, we're number one now, and the revenues of the station have gone up tremendously.

We do 3 hours of news every day, including a 2-hour early evening newscast. There are only four stations in the country that do a 2-hour newscast, and two of them are here in Los Angeles. Our philosophy is that we *do* get 2/3 of our programming each day from the network source. What we have to worry about at KABC-TV then is only 1/3 of our schedule, and with 3 hours occupied by news, you can see what the major focus of the station is. I have the biggest department at the station, with over 100 employees, and the biggest budget as well.

I think there's a strategy to developing a winning news combination, and you can apply it to any station, whether it's an independent, an affiliate or an O and O. You've got to get a successful group of personalities that the audience can identify with. You have to give them some product, so that people know they've gotten the news when they watch these people. And you've got to put some promotion behind it so that people know what you have for them to watch. If you can get that formula together and working for you, you're going to have a winner.

LAWRENCE FRAIBERG: I wouldn't disagree with Dennis about the importance of the news program, but I do think that, if there were a simple formula for getting an audience, there would be many more successful news programs.

My feeling, however, is that at an independent station, the news can't be where our total energies are spent. We try to deal with the *gestalt*—we have a total problem. I think that you first must find your own identity as a station. What kind of news do you want to deliver? Do you want it to be action news, do you want a great deal of humor, or human contact?

I think promotion is important, but I don't think it's how much you spend on promotion. You have to find some reason for those people to tune in, because most people have already sampled you and made their decision. There has to be something else to motivate them to try you again, or to stay with you, and that might be personalities, but it might just as well be another angle, like the miniseries documentaries which have been successful for many different stations across the country. They do three- or four-part stories dealing with sexuality, or psychic phenomena—things that have a lot of "sex appeal" in the broadest sense of that term.

We consider the news very, very important, but it only ranks among three or four other areas in priority, so we must share our energies and our resources. This goes for independent stations generally, and the Metromedia stations in particular. You can't go in for that kind of concentration of energies when you're programming a full television day.

The key to being an effective independent station is counter-programming the network stations. During the 6-8 o'clock time period, when the network O and O's and affiliates are doing news, then independent stations might do children's programming or family entertainment shows. In the afternoon, when network stations are doing soap operas or game shows, then certain stations, including the Metromedia stations, do children's programming. In this way our share of the audience will not peak in prime time, but in the early mornings, or sometimes at 11 o'clock at night. In New York, for example, WNEW-TV (Channel 5) was number one against all other stations when they had *Mary Hartman, Mary Hartman* at 11 o'clock weeknights. They were also number one at 7:30 at night with Carol Burnett against what's called "prime access" programming on the other stations.

We feel that our share of the advertising dollars spent on television isn't enough, but it has been gradually increasing. In New York in 1976, the three independent stations took in 30-35 per cent of the total spot dollars spent in the market, and WNEW had half of that amount.

I think if our ratings ever equal the O and O's and the network affiliates, then it won't make any difference whether we're an independent or an affiliate, because those numbers are really the ones that count. But for now, we have to find devices to try to entice advertisers. One of these devices is research. For instance, we did research which showed what happens when an advertiser buys a particular package of spots—how many people will he reach for his advertising dollar, and how often will he reach a viewer in a given period of time. We sell in competition against the larger numbers of viewers found on the O and O's and the affiliates, and we do this by talking about the "cost per thousand"—the amount of money spent for each thousand viewers reached. We have verified with research data that the cost per thousand can be more efficient with our independent stations.

We will also try to sell in tandem with another station. For example, Channel 5 in New York will go to an advertiser and say, "If you have $100 to spend, why don't you spend $60 on WABC-TV and spend the other $40 with us," and we'll show them the figures on how that will benefit them as opposed to dividing their dollars between two network O and O's.

Besides all the difficulties independent stations encounter in improving their ratings, there is an additional problem—they have not been perceived as being as good a value, rating point for rating point, as the network affiliates. This has to do with bias. The advertiser buys on a cost per rating point basis, and there's a different cost per rating point in different markets. Let's assume that in Los Angeles the cost per rating point at KABC-TV, a network O and O, is $100. Now, for the Metromedia station in L.A.— that's Channel 11—they might pay $50 a rating point. Why? After all, each rating point represents the same number of viewers, whether it's a point on a network station's rating or a point for an independent.

Well, the buyers try to find a rationalization. Their job, of course, is to buy advertising time as cheaply as possible. They'll use any leverage, any device they can, to negotiate a spot package as cheaply as possible. Our job is to get as much as we can, but certainly we feel it's unfair for them to spend twice as much for the same number of viewers on a network affiliate. Some of their stated reasons are: "You don't have the same kind of audience. You have a second-class audience—they're not made up of the same kind of people." "What do you mean by that?" we would ask. "Well, you know what we mean by that."

In order to answer that attitude a study was made, an Arbitron reevaluation study. Arbitron, using a random sample very scientifically worked out, proved that the independents throughout the country and the network affiliates and the O and O's have essentially the same audience. Quantitatively, it's all in that study. Advertisers will see these findings and still resist it. So the bias is not an intellectual problem. It's an emotional problem, an attitude. It has to do with how they perceive you, how they feel about you. It's like a racial bias—no one can justify it intellectually, it's emotional. The only way to overcome it is the same way it was created. You try to make them see you differently, make them perceive you differently, alter your image. This was taken on as a test case in New York. We did a research project, spending a lot of money to find out if indeed there was a bias, as we believed. We did research through focus groups, through questionnaires, and the data showed that there indisputably was this bias.

We then put together a presentation, which later led into an advertising campaign. This presentation was not done on the air—it wasn't shown to viewers. This was shown to advertising agents and to clients, and it was the kickoff for a very expensive, very elaborate campaign. We budgeted approximately $1 million for this campaign, and we spent about $1.5 million.

The filmed presentation which we screened for these people stresses the sense of a New York community, and the independent local station, WNEW-TV, as an integral part of that community. From the news angle, the point is made that WNEW-TV reporters are New Yorkers, not imported talent from other cities. These are people who know the city and how to find out what's really going on there. The object is to show an emotional attachment to the station on the part of the viewing audience—this is their station, their kind of people, not just one more corner of an undifferentiated mass U.S. audience.

Our research showed that WNEW-TV is perceived as an alternative to the network stations, not just as another independent, and this presentation brings that point home dramatically, with interviews, snappy clips from the programming and E.G. Marshall's authoritative voice doing the narration. All very high class, all calculated to overcome that bias against WNEW-TV as an independent station. The hard factual portion of the film deals with a massive media campaign which we were about to undertake, including radio and television commercials, ads in magazines and newspapers, all to remind the New York audience that WNEW television offers a choice, an alternative to network programming.

We followed up the presentation with a mailer sent to all the advertisers, basically a repetition of the points made in the filmed presentation. These brochures were extremely expensive—$25 apiece—but they are worth it if they help to create a positive image for the station. And further research at the end of the year showed that we had indeed succeeded in altering the perception of a percentage of those people.

We think that it will take several years to make this a widespread change of attitude, and we're still in the process of doing it. But we have demonstrated that it can be done.

DENNIS SWANSON: At KABC we have an image problem of our own, which is kind of the flip side of Metromedia's problem. Our image as a station is strongly tied to the image of the network, and specifically in the news area, that causes problems. When you talk about national news, you would place CBS as number one, NBC would be number two, and ABC is a poor third. We have set up Eyewitness News as our local station product, and that's what we promote. When we answer the phones in my news department we do not say KABC, or ABC News. That's because we've built a news image in this town on Eyewitness News, Channel 7 news.

We have carved out a particular segment of the audience, based on a philosophy that there's more to life than news, sports and weather. We do more public service material in our newscasts, we're not as institutional as the other two network stations that we compete with, and that has been highly successful for us, because people kind of think of us as on their side.

We tend to dominate the younger audience in Los Angeles, the 18-to-49-year-old women demographic group. That presents a problem, despite the general desirability of that young group to advertisers. Here in Los Angeles we're competing against the outdoors. You've got a million people going to the beach every weekend. For us to get a decent rating on a Friday night at 6 o'clock, we could get naked dancing girls and it still wouldn't help. Our crowd's gone— they've gone to the beach, the mountains, the desert. It's a more mobile group, because our style of news is a personality-oriented newscast, and that's the kind of hang-loose viewer who will watch us.

Children's Television

Margaret Loesch

Joe Barbera

Margaret Loesch is currently Director, Children's Programming for NBC Television Entertainment, responsible for the development, creative supervision and production of the network's Saturday morning series. She came to NBC in 1975 as Manager, Children's Programs.

Loesch began her television career at ABC-TV in 1971, starting out as a film clerk and working up to the position of Production Manager in the Creative Services Division. Before entering the television business she worked in the stock brokerage field. Her educational background includes a B.A. degree in pre-law and political science and graduate studies in government and international law.

As President of Hanna-Barbera Productions, Joseph R. Barbera directs operations of a highly diversified Hollywood studio. While currently involved increasingly in the production of live-action films and development of theme amusement parks, the studio has roots that lie in the production of animated cartoons. After attending New York University and the American Institute of Banking, Barbera began a career in the world of finance. At the same time he started submitting cartoons he drew in his spare time to major magazines, where he found considerable success. He soon became a regular contributor and, abandoning banking for a career as a cartoonist, became particularly interested and skilled in the techniques of film animation.

This skill led him to Hollywood and the animation department of Metro-Goldwyn-Mayer Studios, where in June, 1937, he met his lifelong partner and business associate, William Hanna. The two men developed the popular Tom and Jerry cartoon series, which they produced for MGM for the next 20 years. Finally, in 1957, they left MGM to form Hanna-Barbera Productions and exploit a niche which the major studios had not yet filled—producing low-cost cartoons for television. Their long string of successes has included such well-known shows as *Yogi Bear*, *Huckleberry Hound*, *The Jetsons* and *The Flintstones*, and the current *Scooby-Doo*.

SONNY FOX: We are going to deal here with a very special area of television, children's television. Let me hasten to explain something that's probably quite obvious, but it gets confused in the arguing about television for children. We are not speaking necessarily about the television children watch, but that television which is specifically *designed* for children to watch. A great deal of the unhappiness about what children watch is really directed at programming which never was designed for children. Indeed, we find children watching through some very strange hours of the night, much later than you think. The idea that you can insulate children from substantial damage by isolating one hour from 8 o'clock to 9 o'clock, calling it the "Family Hour" and keeping untoward sex and violence out of that hour is probably a chimera, because there are probably almost as many kids at the set at 9 o'clock as there are at 8 o'clock.

By and large, the times we are talking about as children's programming are 8 in the morning for most of the country (it's 7 in the morning on the west coast) until sometime between 12:30 and 2 o'clock, depending on the individual network, on Saturdays. In addition to that, there are the afterschool specials, of which NBC does four new ones and four repeats a year, on a monthly basis, on the first Tuesday. ABC presents seven new ones and seven repeats on a twice monthly basis through the season. CBS programs a half hour once a month.

To start out we should ask Margaret Loesch to explain where NBC is currently, at the end of March, in the process of buying, scheduling and developing shows.

MARGARET LOESCH: We have just announced our schedule of new shows that start in September, on September 9 to be exact. Traditionally at this time of year we have just closed our development season and are starting production on the shows we selected. Children's programming at the network level has been very seasonal, with commitments made to programs for the full year. As a matter of fact, only in the past two years have we gotten into mid-season new production for mid-season shows. Traditionally the schedule has been that from September through February we develop product with the major houses, like Hanna-Barbera. At the end of February or the first of March, we make up our schedule by selecting from the shows we have developed. From that point on, March through September, we are in production. Then the cycle repeats itself. As I mentioned, the networks have just recently started developing new product for mid-season, so the very strict season that I've just explained is changing. As a matter of fact, we are just now negotiating a deal with Hanna-Barbera to start developing a mid-season series for January or February of next year, so that during the summer, while attending to production of current shows, we will also be developing a series for mid-season.

The Birth of "Godzilla"

SONNY FOX: Joe, I know that your company right now is dealing with ABC and NBC. Are you doing anything for CBS this year?

JOE BARBERA: Yes, we have an hour, and we've resurrected *Popeye*. And why do you resurrect *Popeye*? Saleswise, when you walk in somewhere and say, "I have a new series with a green elephant, and I have *Popeye*," they always seem to go for *Popeye*. They're not going to buy anything new if they can help it. It's show biz. For instance, you're thinking what NBC might want, what CBS might want, and you try to come up with something, and you wake up one morning and think, "God, no one has touched *Godzilla.*" Now, you know it would have been death to mention that 3 or 4 years ago. So I call the man who controls the property, and he's involved with Toho in Japan, and I say what are you doing with it? Why don't we take a run at it for Saturday morning?

You have to keep thinking of these kinds of projects, or someone else will beat you to it. Then, after you think of it, you know something like *Godzilla* is going to be a tough sell. So, how do you handle it? First, he's a superhero, second he's going to help the environment, third he's going to come to rescue us from incredible menaces that are going to ruin the earth, and fourth we include a new character that's a small version of him, which was once used in the movies, called Godzooky. He's an eager beaver who wants to become like Godzilla who blows out smoke and flame, but when Godzooky tries he just comes out with smoke rings. In this way I'm getting around the programs and practices departments who will stop everything you do that's violent. I can assure you that we've been on a nonviolent kick for over 10 years, no matter what anybody says. If there's any violence, it's an old product. Certainly nothing we've done over the last 10 years.

SONNY FOX: Let's stay with *Godzilla*, and track that project as a case history of how something gets put on the air. I can start the story, because it actually started when many, many months ago a man named Hank Saperstein, who owns UPA, came to me in New York and said, "What about doing *Godzilla* on Saturday morning." Now, I remember *Godzilla* from the Japanese movie *Godzilla*, and I said, "You've got to be kidding, Hank—they'll hang me by my thumbs. *Godzilla* is everything that everybody says we can't do for kids any longer, and if I walked into my management and said, 'Let's do *Godzilla*,' they'd throw me out." So Hank picked up his bag and walked out. Many moons later, Joe says, "What about *Godzilla*?" But Joe then said, "Wait a minute, what if we turn him around and make him a hero figure," as Joe just explained. So then I began to see a way around it, and I said, "Terrific." I wanted it, because my instinct was that *Godzilla* would be a very big hit, and God knows, NBC was looking for big hits. Anyway, I left NBC.

MARGARET LOESCH: And Sonny's instincts were right. It has all the makings of a big hit. We at NBC actually got into a bidding war with ABC, both wanting the property, which happens very often. Normally, on Saturday morning television NBC will purchase 13 episodes of a project over one year and repeat it so it plays a total of four times, totalling 52 weeks. We are an exception—ABC and CBS purchase product over 2 years for their Saturday morning schedule. They buy 16 episodes and run them over 2 years, playing them each six times. The reason for the difference is that ABC and CBS have Sunday morning network programming which allows them to play off shows that cannot function competitively on Saturday morning. NBC does not have that. Our Sunday morning belongs to network sports and to local affiliated stations. So we have only Saturday morning, and if something doesn't work for us competitively we don't have somewhere else to play it off. Therefore, we make one-year deals.

The way NBC worked out acquiring *Godzilla* from Hanna-Barbera was simply that we broke tradition and made a 2-year deal, buying 26 episodes to play over 2 years. This also is an extraordinary situation because it is a deal that was worked out first with Hanna-Barbera, Toho Japan, and UPA, who work in conjunction with Toho. We have our contract with Hanna-Barbera.

We have had, surprisingly enough, very few problems with our broadcast standards department, which monitors everything we do in programming, thanks to the twist in the concept, making Godzilla a hero. We also have a social science advisory panel that was commissioned a number of years ago who review all of our material, and they also have had no problems. We have still been able to create a very exciting episode, and the Godzooky character that Joe described serves very well for comic relief. The show has a great deal of comedy in it.

JOE BARBERA: This is an hour show, but Godzilla himself

comes up in only one segment, and then he only comes in at the end, when the big deal and the big problem comes up. We don't stay with him and firebreathing. We have a segment with a Cousteau-type hero travelling around the world, environmentally trying to straighten things out, and he's a humorous character. That with *Godzilla* makes up a half hour of an hour segment. We still are working on two more segments that will make up the entire hour.

MARGARET LOESCH: We're going to sandwich two 11-minute segments on either side. Right now we're trying to decide on those segments.

Two weeks ago we announced that we acquired the show, finally put it on our schedule, which is now complete. We are now at the script-writing stage, and have started the storyboards, and Joe is working very diligently to get the kind of look we want. We have just finished auditioning voices and recording the first show. So you can see where our schedule stands in March. It's a very time-consuming process, and we probably will be producing this series right on through November.

SONNY FOX: Joe, what's the lead time? When is the first time you will see a first finished episode of *Godzilla*?

JOE BARBERA: Well, if it's normal, it will be one minute before it goes on the air. You get on the plane at 11 o'clock the night before and fly it into New York. That's how desperate these things are.

But actually, yesterday we were testing voices for Godzilla and Godzooky. You have to see this to believe it, because here's Ted Cassidy, who's a huge giant of a man, and he's roaring, and then Don Messig is answering him as the small one. We have to get personality and fun into it. Now we live and die on the pick of the voices. As Margaret knows, I've been running up and down that studio dumping voices, cancelling voices, picking up new voices. We look desperately for new voices. I was trying to get a personality with the little one, and I tried to get some words out of him. Well, he came out sounding like an idiot. So now we got two actors in a room and they began to communicate with sounds, and we were getting some laughs out of it.

Deficit Financing

SONNY FOX: What are the economics inside Hanna-Barbera on a show like *Godzilla*? NBC pays what seems like a lot of money. You apparently have to give some of it to Toho.

JOE BARBERA: We have a license fee with Toho. We have a man here who's kind of guarding the property, so we meet with him once in a while, as little as possible. He reads the script and tries to give us some advice. Then we pay a license fee to him, and there are merchandising rights which sometimes he controls and sometimes we control. Then there are foreign distribution rights which are split sometimes. It's a lot of juggling around to see who ends up with any money. We happened to get the highest fee in the history of Saturday morning programming on this property. To explain that you've got to go back to my original statement. It's only there because *Godzilla* is a known commodity. An article in the *Wall Street Journal* 6 or 8 months ago mentioned that Godzilla has never lost money in the theatre. Every single picture that's been made, and at this moment they're starting to make another one and have asked us to participate. So then when I say, "Let's do *Godzilla*," I find a receptive ear.

SONNY FOX: There's another factor in the high payment NBC is willing to give you, Joe, and that's your own past success, because your shows were so strong on ABC, and of course Hanna-Barbera down through the years has such a fine reputation, that they would be much more willing to reach to meet your fee than they would for somebody else who doesn't have that track record. That might make a difference of $5,000 an episode on licensing fee, to give you an arbitrary figure. There are factors in there that the network would consider, that you have to be aware of if you're selling a show or making a deal with a network. There really is a reward for success over a long period of time, a proven track record.

But getting back to the economics. Most of the time when you do a show for NBC or CBS or ABC whether it's a half-hour or an hour, whether the license fee is $100,000 or

$500,000, whatever it is, you actually lose money, do you not, on producing that show?

JOE BARBERA: Quite often, yes. One of the reasons that we hang in there with red deficit figures, which no one seems to understand, is that we have built up a backlog over the years which is out working for us today, still being sold in various ways. Almost everything we've made over the last 20 years is at this point running somewhere for some reason, for 15¢ here or 2¢ over there. That's what keeps us going. It's the income from merchandise, any income from foreign sources, and the fact that sometimes a network will buy a rerun of one of our products to fill a hole when something else drops out of the schedule.

Once you have your studio set up you're ahead of the game. I can give you some quick figures. When Warner Brothers tried to start their own studio, about eight or ten years ago, they figured out it would take them about $800,000 to crank up, which means people, staff, bookkeepers, unions and all that. If they sold one show they would be in the red; if they sold two shows, they would break even. That's starting from scratch.

I have to admit, frankly, that in 1957 my partner and I started completely from scratch with $4,000 each. You're going to ask how did you manage that, and it's quite a story, but we did. We were lucky. We did *Tom and Jerry* cartoons until 1957, and I'll give you my tale of woe on that count. Bill Hanna and I did every single one of them, wrote them, produced them, directed them and finally left MGM. They're still running them. They might make $20 million this year in syndication with the same things that we did, and Bill and I don't get one dime out of that. It's like movie stars who see their pictures at night on the Late Show and they don't get a dime and they're starving.

From that we went into our own studio and hit with *Huckleberry Hound* and *Yogi Bear,* which by the way we're doing again for NBC as the *Yogi Space Race.* Those characters are like owning a TV star. If we keep them going, we're lucky. That's how we can exist.

SONNY FOX: Put another way, in terms of the deficit financing, if it costs Hanna-Barbera $110,000 to produce a half-hour show, and they're getting $100,000 from NBC, they are deficit financing $10,000 a week. On the other hand, they're using $100,000 of NBC's money to create a property that will be owned after the first year, unless options are exercised, by Hanna-Barbera. So for $10,000 a week, or $130,000 all together, they have now created equity for themselves. They have created a property, and animation has an incredible shelf life, doesn't it?

JOE BARBERA: Yes, but there's one more thing to keep in mind. If I do a show like *Popeye* or *Godzilla*, we don't own that outright. We participate with many partners. On *Popeye* we have the toughest partners in the world, King Features. That's why we prefer to do something original.

Scooby-Doo, for instance, happened to become a gold mine. Don't ask me how it happened—nobody figured the dog would become a star, but he has. He's been on the air over nine years, and he'll be on about four more years. Like anyone who's creative, you'd much rather get your own character out there and make a success out of it, but the kids will keep turning to Superman, Batman, whatever.

You never know when an idea will present itself, and you have to be open to those ideas all the time. I met a fellow at breakfast one time who was the head of King Features and I said to him, "What are you doing with *Popeye*?" and out of that came the deal to do *Popeye*. They weren't doing anything, but they were thinking of doing something. *Godzilla* was just waking up one morning with an idea and calling the guy. We work with DC Comics, and one day I picked up a cover and there were two long rows of characters, on one side were the superheroes, and on the other side were all their super enemies. Each one had a special enemy, like Superman has Lex Luthor, Batwoman has Cheetah, and so forth. And I looked at it, because I'd been watching the *Battle of the Network Superstars* and all that on television, and I said, "Holy mackerel—*The Battle of the Superheroes!*" So I made a call and we had a deal on it in about two seconds. Now we have to negotiate with that company, and they get a nice piece of that. But you're only as good as your ideas.

Training Animators

SONNY FOX: True, ideas and execution. Let's go to execution. You made a deal with NBC to deliver a property called *Godzilla*. Do you have to go out and hire people now, in addition to the people you have in house to do this? How much of an operation is involved? I mean, your organization works like an accordion, doesn't it?

JOE BARBERA: We've instituted some policies in the last two years which have changed that a little. I'll backtrack a bit. When we were doing motion picture cartoons, we worked a full year. It was a steady job, Disney was steady, Warner's was steady, and the people who were being trained were just super people. Suddenly it all stopped, because they stopped making movies and started going into television, which was seasonal. So for 18 or 20 years we had almost no new talent. We couldn't train anybody, because if a young aspiring artist walked in the door we would say terrific, look at these samples—but we've got to tell you, you only have a 6-month job, and then we have to lay you off.

What we've done lately, and with the fact that you have mid-season buys now, we're able to keep people working all year. We used to fire everybody just before Christmas. But now we're making animated features, which is a company policy, through which we are training people. We have a school. We have 134 people that we put on and trained last year. We're now training writers, and we're the only ones who have started to do that seriously, because the talent has just been disappearing. When people came into our studio, they'd see our staff with seeing eye dogs, in wheelchairs. We've been having a terrible time producing the amount of work that we are lucky enough to sell. Selling it, creating it and thinking of it is one thing. Now comes that big production machine which has to be going, and you have to try to keep the quality up.

In the last few years we've overcome some very strong resistance and started to do some of the animation abroad. We opened a studio in Australia, started right from scratch, and then, to be smart about it, we sold half of it to an Australian company because it's very tough to operate down

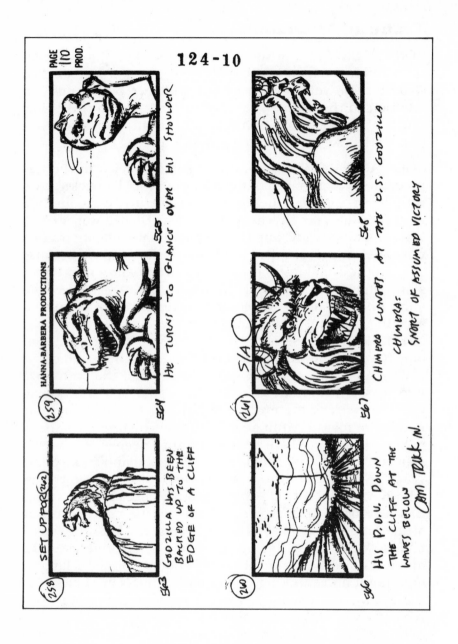

A story board from the Hanna-Barbera *Godzilla* cartoon show.

there without an Australian partner. That is normal all over the world, they all move in as your partner.

We did train a group, and they're excellent animators now. They're technically super. The one thing they can't do is create ideas that will sell in America. That's their ambition. They want to come in and cut our throats (this is our own company). And we have a group in Spain that is very good. We have a group in Mexico, another in Yugoslavia, and we're talking to a group in Taiwan that's very good, but mainly for one thing—animation. All the stories, the layouts, the storyboards, the creation of the characters, the voices, are all started in our studio. None of them seem to have the knack, fortunately for us, of creating subjects that will sell to networks.

Foreign Animation

SONNY FOX: The foreign animation work has gotten so good that last year CBS insisted that some of their shows be done abroad, and NBC got to that point too. I was very suspicious of it. There had been, some years ago, an attempt to do it there, and it was not done well. The control was a big problem. So we proceeded gingerly and insisted on having some tests made abroad, but they turned out to be good. As a matter of fact, Margaret was convinced that the stuff coming from abroad was superior to what we were getting done here.

MARGARET LOESCH: It was. We found that during the crunch, the part of the season when the pool of talent is being strained, certain types of comedy animation are done better out of the country, and NBC has requested that some of our animation be done in Spain and Australia. As a matter of fact, we've approved some of it to go to Mexico City to try it out and see how it looks.

Animation on a TV Budget

SONNY FOX: On page 144 you'll see a Hanna-Barbera storyboard for *Godzilla*. The normal way of doing a storyboard is to tell a story frame by frame, writing out the script and including enough pictures so you can follow the script and get a sense of how the animators intend to direct the sequence. Then we can look and see if there's enough action, if there's too much talk, or too many head shots. Now we may ask for things the animators can't give us. It may simply be too expensive to do what we want to do for the price. There's always that little argument that goes on between the network and the animator. The network says, "Why can't we have more of this and do this a different way?" and the animator says, "Because you're not paying us enough."

MARGARET LOESCH: We work very closely with the animators, going over original storyboards and revised storyboards until we agree on the way a sequence is to be done. To give you an example, in a fight sequence between Godzilla and the firebird, which is a pterodactyl type of monster, we have a staging of a scene where Godzilla and his adversary are clutching and fighting, and there's a shot indicated where they fall offscreen down into the water. Now, my note was that I'd like to see a dramatic shot. I'd like to see it from above, and look down and watch them plunging into the water instead of having them just slide off camera, and in the next shot show a splash of water. Those are the types of things I'm always looking for, and that's one of my notes on this board.

SONNY FOX: All right, let's use that example. Joe, you're the practical guy, because this means dollars and cents for you. What is the difference to you between two monsters falling offscreen and dropping, and watching them from above getting smaller as they fall?

JOE BARBERA: It so happens, I think I told the animators to do the scene this way for that very reason, because they fall offscreen this way. Margaret would like to watch them fall, which means we have to cut back, do a longer shot to see them fall, and they're going to hit into the water and splash. Well that means a lot of drawings of very complicated prehistoric characters, and a lot of water animation. You may have seen slow motion films of a drop of water hitting. First it hits, then it begins to come up, and then another drop comes out, and then it hits, and then it bubbles, and then it spreads. That's 8 billion drawings. That's one of the things that gave Disney some of his biggest problems in his time, by the way. He used to think that that was important. Well we have found out that that isn't important, but Margaret would not concede that. My effect would be that they drop off, there would be a tremendous splash, and then we would cut to the water bubbling and boiling and steam coming up—we'd give you a terrific effect. But she's the boss!

MARGARET LOESCH: You see, this is exactly the type of thing we have to concern ourselves with. He has to worry both dramatically and economically about what it is we're requesting, and we worry about it dramatically. One of the things that we can do which does not require a great deal more animation, and it's something that Joe is a master at doing, is drawing a scene so that it's dramatically staged. It's very exciting, even if it is a still shot and there is no animation in it, if the camera angle is a dramatic camera angle. We've been discussing *Godzilla*. Instead of using a shot of Godzilla just coming up out of the water in a long shot, we would have the shot designed so that it will look as if the camera were under him, looking up, or, as another possibility, as if the camera were over his shoulder looking down into the water. Designing a board and staging it that way does not necessarily require more animation, but more imagination, and Joe gives us those touches.

JOE BARBERA: More imagination, and a certain specialized type of artist, of which possibly on the west coast there are two or three, and in New York maybe three. That's the difficulty with this type of product. We happen to have a super artist for *Godzilla*, but I just wanted to make one thing clear. We are talking about Saturday morning or semi-limited type of animation and not theatrical animation. If you do theatrical animation you're going to talk about five or six times the money. So that's the difference. You have a budget, and you have to stay within it, and you have to do footwork, and tricks to come out with an effect. That's where we started 20 odd years ago with what you call "limited animation," and without it we would never have gotten to television.

SONNY FOX: It's the kind of animation, however, that many critics will refer to as an example of the paucity of production values available to kids. Again, you're looking here at what has become a leitmotif throughout these discussions, that you can start talking about creativity and almost always end up talking about the limitations imposed by the role of the dollar. The money finally shapes what it is that comes out of the tube. It determines the kind of animation that Hanna-Barbera can do. Not what they're capable of doing—not what artistically they can do, not what they're going to do when they film a feature-length theatrical cartoon like *Heidi*, but what can be delivered to the network at the license fee they pay, even when the producers deficit finance that. You've heard it from Alan Landsburg, from Paul Klein, and the others—the dollar is always in there. And the answer to the critics of "limited animation" is that it's the best you can do for the money the networks are giving you.

JOE BARBERA: Well, I have another answer. It doesn't mean a darn thing to the kids, whether I put 40,000 drawings in or 400 or 4,000 as long as the entertainment is up there. They're not aware that when a man is running, his foot hits the ground, and his knee bends, and his muscle quivers, and his hair jumps. That doesn't mean a thing to them, really. On Saturday morning you must give them action and entertainment, and noise. There's no doubt that a lot of people say, "Oh, look at that junk," and "We could do better," and we've suffered, we've burned on all that stuff. We did a show called *These Are the Days* for which we received nothing but kudos, handshakes and pats on the head from all the organizations. It was a family show like *The Waltons*, and the kids are switching to *Batman*. We got all the good notices, but we didn't get the rating points. And if you don't get the rating points, all that work goes down the tubes—goodbye—cancel the show.

The NBC Schedule

MARGARET LOESCH: NBC got the best notices in 1978, and we were number 3 in the ratings. In the new schedule we have just arrived at, we are trying to boost our rating points by going after name characters, characters with marquee value. Any new characters we are trying to create we will sandwich between the names that are already familiar. I'll run through our schedule and you'll see exactly what I'm talking about.

We bought 90 minutes of *Yogi Bear*, a potpourri of cartoons from Hanna-Barbera. Included in that are characters like Jabberjaw, Huckleberry Hound and, of course, Yogi Bear. We have also, within that 90 minutes, created some new characters that will have their own cartoons. Obviously, the attempt here is to create and to ride on the coattails of established strength to create potential spinoffs. We also purchased an hour show called *The Godzilla Power Hour* from Hanna-Barbera studios. We also bought, from another studio, *The Fantastic Four* which Marvel Comics publishes. *The Fantastic Four* has wide distribution as a comic book. It will now be going into animation. We bought a half-hour show from another studio, Filmation. It stars Broom-hilda, Emmylou, Nancy and Sluggo, The Captain and the Kids, Alley Oop—all comic strip characters that have a name value, creating strength already going in. We bought an hour from Sid and Marty Krofft which will be a live action show, and we are right now seeking some singing group with strong name value to host that variety hour.

So you can see that we did not use new characters and concepts we had developed at NBC. Many were excellent concepts, very strong shows. In live action we developed a *Young Davy Crockett* series, a new version of *Flipper*, and others. In animation we developed much more, particularly with Joe, but what we finally decided to go with were things that had marquee value and established strength. We tried to update them and make them look contemporary. For instance, Yogi Bear will be in a new format. Huckleberry Hound will be in a new format. Godzilla will be in animation, a format he's never been in. So, it was with that in mind that we selected the characters and shows that we have.

The procedure at NBC is that once we in the programming area have established the shows that we want to go with and come up with a recommended schedule, we then have to do the same sales pitch to our management that Joe Barbera has to do to us when he's selling a show. I went to New York and made presentations to the upper management, to the head of programming. Once he agreed and made his recommendations and had his input on the schedule, then we turn around and go to the top echelon of NBC management, meaning the president of network.

We also have to work with the research department, which always has their input, and they had done some testing. We are in a marketing world, I admit. We are trying to market a product, so I don't think that the research department is completely full of hot air. However, it's very tough for me as a creative individual to deal with people who make programming decisions based on numbers. It doesn't help us come up with a Godzilla cartoon, but we do have to contend with research at the network.

We have also to deal with sales. The sales department, of course, has to deal with the potential sponsors. They must be enthusiastic about our schedule, and they must feel that they can sell the scheduled shows. As an example, *Godzilla* was a big hit with the NBC sales department because they felt that they had something they knew they could sell on the streets. They knew they could sell this to McDonald's and Kelloggs and Mattel and Keebler Cookies.

Once we have had our meetings with the upper echelon of NBC, sales, research, broadcast standards (we always go through our potential shows with broadcast standards to make sure that we do not have what they think are major problems), then we lock into our schedule. We make our announcements and contact the producers to tell them that we are picking up the shows. The licensing fees for the shows have already been established. The deals have been made. During development season, when we develop a series with anybody, and we give them X amount of dollars to give us a script and a storyboard and a bible on a series, we also, at the same time, make a deal so that if we select the series, the pricing has been established. So that there are no business affairs to take care of when we do decide to take that next step.

The next step on the network agenda is to go into conference with the producers, assign the writers to the shows and start the creative processes. Last week I was on a sales junket to Chicago and New York where I, as a courtesy, to boost the sagging morale of NBC in the sales market, made presentations of our schedule to the various agencies and their clients, in particular Kelloggs and McDonald's, which are two of the biggest buyers of Saturday morning commercial time. So you can see that taking something Joe and I decide together is a great idea, someone somewhere along the way is going to want their thumbprint on that product because of the various steps I've just explained that we have to go through. Someone at one of those meetings will say, "Wait a minute, why don't you do it like this?"

SONNY FOX: There's a lot at stake. An enormous risk will be taken on the basis of recommendations from Margaret and Mike Brockman, the vice president for daytime and children's programming. You've got 4 1/2 hours of new programming there. You're probably in for about $13-14 million expended on new product.

MARGARET LOESCH: It's $14,200,000.

SONNY FOX: That's over $14 million for just 5 hours of Saturday morning. Now the sales people and the marketing people have to take that and figure out how to sell it in order to make it pay for NBC. They try to lock advertisers in for 52 weeks. They go to the big buyers first, then fill in from that point. It's easy to sell out the last quarter of the year, because everyone wants to get on before Christmas. It's the other parts of the year that are a little bit more difficult.

You sell off a projected share. That's important to know. The research department will look at Margaret's schedule and then assign a projected share through some mystical method that totally escapes me. One time I remember, Margaret and I were in a meeting, and three researchers were down at the end of a long conference table, and they were coming up with numbers. I suddenly looked at them and realized who they were—they were the three witches from *Macbeth*, and they were stirring their cauldron of numbers. They'd pull out a figure and, if that didn't work, back they went and they'd pull out another. Luckily Margaret knew more about the numbers they were talking about than the research department, because she does that statistical thing better than anybody I know. She would say no, that's not right, and they would go back to their cauldron and say, "OK, how about this one?"

Anyway, they come up with an average for Saturday morning. Last year they sold off a 26 share projection. That is, some shows they projected at 18, some at 30, and it all averaged out to 26. Now it's got to be as high as they can make it to sell at a good price, but not so high that the people won't buy it at that price because they don't believe you. CBS made a basic error in 1977. They went out and sold a 34 share projection and they didn't sell out because they were selling at such a high price based on a 34 share that a lot of sponsors said, "Wait a minute, that doesn't make any sense to us." NBC sold out very quickly, before anybody else. CBS waited, and as a result they had to start cutting rate cards in order to get into a sold-out position. So, it's a very cute balancing act that the sales people do. Just out of curiosity, what share did they project this year?

MARGARET LOESCH: 26.4. The year before it was 27, but they came down. It is a mystery to me.

SONNY FOX: It figures, since NBC has been averaging considerably less than a 26.4. Sponsors wouldn't believe 30, they wouldn't believe 32, but 26.4 based on *Godzilla*, on Hanna-Barbera, on the very big superhero orientation, on the fact that these are not soft shows, not noble shows, just straight-out bombshell-type shows, that 26.4 would probably be believable, and they will sell heavily. I would say that within the next four to six weeks, most of the sales will be finished.

Money for Production

QUESTION: Hanna-Barbera is now owned by Taft Broadcasting. How does that work to your advantage?

JOE BARBERA: Actually, if you have a business of your own, and it's working fairly well, it's very hard to get any money out of it unless you sell it and get a capital gain. It's an odd statement, but that's the only way for a small company like us. DePatie-Freleng is trying to do the same thing. Eventually, if you build an asset like we did with this company, you find another company that is looking to diversify. We had three people interested in it: Universal, Columbia and Taft Broadcasting, which eventually bought us. In a sense, it gives you a money base that takes care of some of your deficit. They have money, and will back you in projects. For instance, Taft is backing us in making three or four motion picture features that we're working on now. Without Taft putting up the money, we wouldn't be doing them, and it took 3 years to talk Taft into that expenditure.

I'll tell you something strange on the Wall Street side of it. If our company announced that we were going into motion picture production, the analysts on Wall Street would be terrified, because it's possibly the trickiest business in the world, and you can lose everything. But if we announced we were doing an animated feature, they'd be delighted. Now, what is the reason for that? If you do an animated feature that works, it's a lifetime asset. It will run forever. That's how Walt Disney built up his library, and his studio's still holding most of those films. You have to remember there were times when Walt Disney was so broke, after *Fantasia* and some of the others, that he hocked his 16 mm. library for a million dollars to raise money. That's how tough things were. They were received well, they were masterpieces, but they just didn't make it financially on the first time around. Since then, they keep making more on reruns. Every time they bring *Fantasia* out they make more money than when Walt first started. So that's how I convinced Taft Broadcasting, and that's why Wall Street backs us in that kind of thinking. You have a lifetime asset if you make an animated feature.

SONNY FOX: You see, in television you don't go into production until you've made the sale. You know you've got the
money coming in every week. If it's a bomb, you've lost a few
bucks, but you know you're going to get the bills paid. With
a motion picture, all the money is spent up front and when
the picture's in the can, then you go out and find out if
you've got a sale. That's what's so scary about the motion
picture business, but of course Joe's absolutely right. If it's
a winner, years from now that animated film will be as valid
as it is today, and will still be making money at the box
office.

Advertising in Children's Television

QUESTION: Can you give us your opinion on the con-
troversy involving any advertising to children on network
television?

MARGARET LOESCH: That's a loaded question, and I'll tell
you why. I don't have a crystal ball. I can't predict the future.
I will say this much—if advertising is taken away from the
Saturday morning market, there will be no more Saturday
morning programming. That is a fact of life. However, it is
going to be a major battle that will ensue between the advertisers, the networks, the FTC (Federal Trade Commission)
and the FCC (Federal Communications Commission).
Everyone is gearing up. My personal opinion, and this is not
an official NBC position, is that the court fight is going to be
over First Amendment rights, and I would hate to see that
come to pass. I would hate to see the day when we cannot be
self-regulatory, but would have an outside agency of the
government step in and tell us what to do. I would rather
someone say, "You regulate yourselves," and we will do that.
We have been doing it in programming for years.

The networks have also exercised control over the content
of the commercials which are aired, and the public doesn't
seem to be aware of this. I am sure that in the future this
control will become even stricter. We do screen and regulate
the content of commercials because some years ago,
Carter's Little Liver Pills, if you recall, were found to be
using fraudulent advertising. It was the networks who re-

fused to air these any longer. Now the networks do screen commercials for fraudulent advertising. We are on the line, they go on our airwaves.

JOE BARBERA: If they stop advertising on children's television we're out of business, but I don't see that day coming. I was meeting with the Kelloggs people not long ago and I know that those people are in their labs today creating cereals that absolutely do not have one speck of sugar in them—they're the healthiest things in the world. Later on we were sitting in the breakfast room at our hotel and over there were two kids having cereal, and they got the sugar and put it on themselves. So what's this presweetened cereal problem we're having? Kids are going to put sugar on cereal no matter what you do. It's just an act that's going on with some of these people who are getting a lot of attention. I tell you, the control has got to be in the home. I keep saying it's the parent that's going to run how much sugar you put on something, what you watch on television, and nobody's ever going to change that. I don't like regulation.

SONNY FOX: There is a valid issue here, though, about whether the younger kids should be exposed to advertising at all because of their inability to judge what is and what isn't good.

MARGARET LOESCH: The real problem is that this kind of reasoning has opened up a Pandora's box. Saturday morning programming has far fewer children viewers than 8-9 or 8-10 o'clock evening programming, as Sonny mentioned earlier. The commercials that you find on Saturday morning, Nestle's, as an example, or Mattel, also appear in your 8:00 and 8:30 programming. Where do you draw the line? This is a layman talking now, because I'm not a lawyer, but I see tremendous problems being uncovered. If they try to regulate in one area, why not regulate in another. If the issue is the impression you can make on young children, I would think that what must be done is to stop young children from viewing television, because they're viewing everywhere, not just on Saturday morning.

SONNY FOX: Except that the product advertising on Saturday morning is designed for and aimed at that specific audience, the children. That's why that issue is focused there. One of the most interesting responses to that question, which I find a good one to think about, is that if indeed the Surgeon General or the HEW people or somebody finds that presweetened cereal is really a danger to children, then that product ought not to be allowed to be sold. And if it's not allowed to be sold, then of course there will be no advertising because it simply will not exist any longer. Why tackle it by saying, "Don't let it be advertised on Saturday morning?" If it's a danger to the children, be brave, be bold, and be honest about it and take it off the marketplace. Television tends to be the handy whipping boy. Also, a lot of this clamor is led by newspapers, which of course stand to gain considerable revenue if there are any limitations put on advertising a product on television.

Sports

Don Ohlmeyer

As Executive Producer of NBC Sports, Don Ohlmeyer exercises creative control over all network sports productions. He joined NBC Sports in May, 1977, after nearly 10 years with ABC.

Ohlmeyer began his broadcasting career with ABC as a production assistant in 1967 upon graduation from Notre Dame, where he received a B.A. degree in communication arts. In 1968 he became an Associate Director, in which capacity he served at the Olympic Games in Mexico City. He has played a major role in covering each subsequent Olympics. In 1972 he was Director of Olympic coverage at Munich, at the 1976 Summer Games in Montreal he was Director of ABC's telecasts. He was also Producer/Director of the 1976 Winter Olympics coverage from Innsbruck. When NBC won the rights to televise the 1980 Moscow Olympics, they convinced Ohlmeyer to join them and supervise all aspects of the preparation and presentation of the network's intensive Olympic coverage.

Ohlmeyer has been honored with seven Emmy Awards, including recognition for his achievements in Olympic coverage, his service as producer of ABC's *Monday Night Football* games and ABC's *Wide World of Sports*. He has also been involved in producing non-sports programs, including the 29th Annual Emmy Awards telecast and the "Battle of the Network Stars" show, and continues to function as a program packager for NBC-TV.

Sports programming on television is now a billion dollar business. The most recent figure was that the three networks programmed a total of 1,200 hours of sports a year— that's an awful lot of hours. When you get into September and October you're at the tail end of the baseball season on Saturday, with doubleheader football on Sunday and the World Series on prime time, plus *Monday Night Football.* That's a tremendous amount of sports on television, and for the most part it is doing well.

The sports departments are only now adjusting to the enormity of their operations, and the first one to recognize the scope was ABC. It's a $250 million a year business at each network, and yet when I got to NBC, they only had an executive vice president and a couple of directors. Find me a $250 million company that is run by three people, and has 25 employees. The networks are finally starting to catch up with the fact that this is a major business.

I became involved as executive producer of NBC sports rather reluctantly, because my goals were to get into other things. I had signed on with NBC basically just to do the Olympics. I did the Emmy Awards telecast and some other entertainment shows, and I was very interested in branching out. But it became apparent that the only way to develop an organization capable of covering the 1980 Olympics was to get involved with the day-to-day machinations of NBC sports. NBC sports slogan has been, "The number one network in live sports," which doesn't mean all that much. I think NBC sports was number three, and we've had to work up from there. Television has become so diversified that you can't cover just one area. You can't just do live coverage or do tape, you can't just do edited shows or specials—you've got to do it all. That's the capability we're developing.

In order to do that, you must determine a focus and build an organization to meet those goals. Then you have to be willing to work hard enough to carry the project through.

The Importance of Personalities

What we're developing now is our focus, agreeing on what we want to do, and deciding what people we want on the air. We can't just haphazardly put whoever's available on the air. We have to decide who we are going to develop into "stars"—who is capable of becoming a "star." I think there are two basic reasons ABC has been successful in the minds of the public. First there's the fact that in most areas their production expertise has been far superior to their competition, but, more important, ABC has developed "stars." You can probably name five "star" ABC announcers—Howard Cosell, Jim McKay, Keith Jackson, Don Meredith, Frank Gifford—that everybody who follows sports television knows. Now take NBC or CBS—who can you really name? For NBC there's Curt Gowdy and Dick Enberg—that's about it in terms of "names" on the same level as the ABC people. For CBS you might possibly put Brent Mussberger in that category, or maybe Pat Summerall. The development of sports personalities has had a great deal to do with ABC's acceptance, and it becomes a cyclical effect. The bigger the announcers get, the more people recognize them, become comfortable with them, like them, and the more they watch their shows.

There are a great number of negative things said about Howard Cosell, and yet people tune in to see him. I firmly believe that if you took the same event on two networks at the same time and one of them had Howard Cosell on it and the other had somebody else, the one with Cosell would get a bigger audience, because Howard's the kind of guy people either love or love to dislike. Those that dislike him love to sit there and disagree with him, to sit with a bunch of people and say, "Isn't he stupid?" He's not stupid. In fact, he's probably one of the most brilliant people I've ever worked with. He was editor of the Law Review at New York University, Phi Beta Kappa, with an incredible mind and an unbelievable vocabulary. What we did with *Monday Night Football* was to try and capitalize on the controversial na- ture of Cosell. In the 5 or 6 years I was involved with *Monday Night Football*, we knew that we had a certain number of people who would tune in no matter what we did. This gets back to the question of focus. Your hardcore foot- ball fan is going to watch whenever there's a game on, who- ever is doing the commentary, but that was not enough to make *Monday Night Football* successful. What becomes the focus of the show is to appeal to people *other than* the football fans—you've already got them. That was the genesis of having Cosell and Don Meredith on the football broad- cast, to try to add some entertainment that might appeal to those people who are *occasional* football viewers. Then, in the second year, Frank Gifford was added. Frank is a hand- some, attractive man who men feel comfortable with, and who appeals to women. I think that was the initial reason for his being there, not because of his expertise as a play- by-play announcer.

Those people were put together to create an excitement so that, even if the game got out of hand, and the outcome was obvious, people would stay with it, because they were afraid Cosell or Meredith were going to say something so outrageous that everybody in the office or at school the next day would be talking about it. It's a little bit like watching *The Tonight Show*. If you notice some outrageous people on it, you're afraid to turn it off, because you don't want to be the one guy who didn't see Robert Blake take his pants off. It's a stupid comparison, but people are like that. We did a game once where Los Angeles was beating Philadelphia, 42-3. We still ended up with a 19.9 rating and 33 share, and as the game went on the audience did not drop off that much. It was not because people wanted to see if Philadelphia would get beaten 64-3. It was because it happened to be one of those nights, which happen maybe four or five times a year on *Monday Night Football,* when everybody was rolling. The announcers were outstanding. Howard was crazy, and they were yelling and screaming at one another and arguing, and the people loved it. They just stayed with it *despite* the game.

There were some times in the last year before Meredith left that I felt I was losing my sanity. There were two very volatile personalities involved in that show, and Frank Gifford became almost like my shrink, counseling me on how I could keep these two incredible egos, Don and Howard, from killing each other.

The coup de grace of the whole series came one night when I got into the limousine with Frank and Don. We had two separate limousines because Howard wanted his own. Frank turned to me and said, "We're not going to talk to him tonight."

I said, "What do you mean you're not going to talk to him tonight?" and he said, "Well, he's been saying so many bad things about us being jocks in the press, that we're just not going to talk to him tonight."

I said, "Frank, you're about to go on national television for 3 hours. There are only three of you in the booth. How can you not talk to him?"

And he said, "Well, he can say whatever he wants, but Don and I are going to talk to each other and not to him."

And I said, "Oh, my God," and we talked a little bit more and finally they acquiesced and grudgingly talked to him briefly about two or three times a period. Probably a lot of people at home didn't realize it. Out of live television, which is totally chaotic and insane, comes a very controlled picture and sound presentation. The guy at home has absolutely no concept of what goes on to get that picture and that sound there.

Sports Economics

SONNY FOX: The increase in air time devoted to sports would seem to indicate that there must be more money to be made from sports programming than from some other form, like news or entertainment. What makes sports economically appealing?

DON OHLMEYER: They're not always appealing, even though the ratings for sports are generally high, which raises the amount of money that advertisers can be charged for time in a sports broadcast. Also the demographics of the audience are highly targeted. Add to that the fact that production costs are basically minimal, for the size of the business we're talking about. A *Monday Night Football* game will cost maybe $120,000 below the line, which means everything other than talent. Acquiring broadcast rights is expensive, though. The rights in 1978 for a *Monday Night Football* game cost around $2.5 million per game.

SONNY FOX: That's $2.5 million for a 3-hour program, which comes to about $800,000 an hour for the rights. Add in $120,000 below the line for the whole 3 hours, plus another $100,000 above the line for talent, and it's costing you about $1 million an hour. That's more than you pay for prime time programming, since a 2-hour movie now runs about $1.2 million.

DON OHLMEYER: But there are more commercials in sports. In 3 hours of *Monday Night Football* there will be 22 minutes of commercials, whereas, in prime time there would be about 6 minutes an hour, or 18 minutes total.

The problem is that the rights have gotten out of control. Where sports used to be a very profitable operation, it is no longer, particularly with the new NFL contract. Many people think that the NFL contract will put a lot of sports departments close to the red for the first time in many years. That's part of the reason TV is going more toward the manufactured event. We're still going to cover some of the major sport franchises, but people really have this thirst for sports on television, and there are only so many franchises. If you can create your own franchises and make them successful, you don't have to worry about the enormous rights expense. It's part of the problem with television, in sports and in prime time—you create your own Frankensteins. You devote so much time to the promotion of football that 15 years later you have to pay $650 million for rights because you've been successful. You promote and develop *Charlie's Angels* and make a star out of Farrah Fawcett, who then becomes a problem and leaves the show.

QUESTION: You talk about creating Frankensteins by promoting sports franchises, like the NFL. Is it possible to destroy them as well by overexposing them?

DON OHLMEYER: No question about it. Boxing is the perfect example. Boxing was destroyed back in the 1950's by television, and I think it will be destroyed again through overexposure. Exposure is terrific for sports. Exposure can make a sport, but that exposure needs to be controlled. I think the best manipulator of exposure is NFL Commissioner Pete Rozelle. Every football team owner should put away a million dollar trust fund for Pete because he's done such a fantastic job of nurturing the television exposure of football, and really parlaying it into a contract which is absolutely beyond human credulity. According to the network contract with the NFL each team will get $5.2 million before they open their doors each year. Five years ago a football team was sold for $17 million. If you assume that you've got $5.2 million coming in from television, and let's say your stadium holds 50,000 people, at $10 a seat—that's $500,000 a game times 8 games—that's another $4 million. So now you have a $9.2 million gross, and that's before the concessions, the parking and all the rest of it. Now how many companies would be sold for two times their annual gross? Those football franchises are worth a fortune now.

SONNY FOX: Let me ask you a dumb question. If indeed bidding is getting to the point where the sports departments are going to be losing money, wouldn't it be just as easy not to bid and just put something else on for those hours? Or would that be so damaging to the prestige of the network that it wouldn't be tolerable?

DON OHLMEYER: It's not so much the prestige of the network. The television industry is filled with rampant paranoia, particularly at the corporate decision-making levels. Pete Rozelle was able to extract an inordinate amount of money from the networks because no one was really willing to tell him "no," and the reason these corporate executives were not willing to tell him "no" was because the basis for a fourth network could be professional football.

If you took professional football as a staple on a weekly basis, you could put together a fourth network. There's enough programming available out there. What's missing is the hook, the one program that the public will not be willing to do without. I think a lot of the network executives saw professional football as the hook. You take NBC, running third in the ratings. If they lost their weekend football games, they could potentially lose 50 per cent of their affiliates overnight, at least on Sunday. Then the "4th network" starts putting together a prime time operation. Relationships develop between affiliates and this other network, and that's what the existing networks are afraid of.

We have now reached the point where the price of the broadcast rights for major sports makes coverage close to uneconomical for the networks. When people look back sometime in the future and try to determine what destroyed sports on free television, they will look at the NCAA football contract, where ABC, really out of desperation, paid far more than it made sense for one network to pay. They did it because CBS wanted to split the package with them, there was a bidding war, and finally ABC came through with all the money themselves. They just don't get enough air time to justify the expense. The colleges play on Saturday—they don't play on Sunday. Unless you have games going head to head with one another, there are only so many hours you can program, and ABC agreed to pay $30 million a year for the NCAA package, which was an increase of 80 per cent. Pete Rozelle had already written down the numbers he wanted from the three networks for pro football. As soon as it was announced in the papers that the NCAA had gotten $30 million a year, he tore up all the numbers and added another 50 per cent or so onto the increase that he had already penciled in. So you now have a situation where the next time around, the networks probably will not be able to come up with anywhere near the same kind of increase for any of the major franchises. And so the only ultimate answer is pay television.

The possibilities for pay television are absolutely frightening. Let's take hockey as an example. It's not big around the country, but take New York City alone, which is loaded with hockey nuts. If you wanted to sell all of the Ranger or Islander hockey games to these people, and all it was going to cost them was $1 a game, you could easily sell a million of those subscriptions. On that basis, an 82-game schedule would gross $82 million, for *one team*. Even if you only sell half a million subscriptions, which in a city of that size is very little, you're still talking about an awful lot of money, more than network television can possibly pay.

Pay television is where it's all going to go. Ultimately, there's going to be a confrontation between the sports business establishment and Congress, and it will be determined, probably by the Supreme Court, that Congress does not have the right to tell the NFL or any other sports authority that they must stay on free television because everybody has the "inalienable" right to see the Rams play football. It comes down to a simple, capitalistic fundamental, which is, "If I own this, why are you telling me that I must give it to you for nothing?"

The whole problem with government regulation and the blackout rule came as a result of the Washington Redskins. If the NFL had been smart, they would all have chipped in however many million dollars it would have cost and built a 100,000-seat stadium in Washington so that all of the government officials could buy season tickets. The whole reason "the government" got involved was because nobody in Washington could get tickets to the Redskins games, and once the Redskins started winning, everybody wanted tickets. That's when the whole blackout thing started.

The Olympics

SONNY FOX: Do you think pay television would be in a position to bid for the rights to the 1984 Olympics? Or possibly a consortium or syndicate could be established for the express purpose of bidding for the rights to the Games?

DON OHLMEYER: Well, you have a problem bidding as a consortium unless they change the International Olympic

Committee rules. The IOC rules state that the rights can only be sold to a legitimate national broadcaster. This is where the SATRA thing came up in bidding for the 1980 rights. There was a point in the Olympic negotiations where the three networks said, "We're not interested. We're going to pool, and we'll come back with one bid." The Russians had overplayed their hand a bit, and they got a little scared, so they signed a letter of protocol with an organization called SATRA, which is the Soviet American Trade Organization. Everybody in the industry kind of laughed, because according to the IOC rule they had no right to sign a protocol with SATRA and the IOC would never recognize it.

So unless that rule was changed you couldn't come in as a consortium. However, if there were a fourth network by 1984, then they could bid, and it wouldn't matter that they were doing it as pay television. I don't think the ultimate distribution would matter as far as the IOC is concerned.

SONNY FOX: The bidding for the 1980 Olympics was very interesting. At the last minute NBC went in and secured the rights for a bid of some $86 million. That does not include the cost of producing the shows—that's simply the rights fee for payment to the International Olympic Committee, and to the Russians for the equipment which they would have to use to produce this broadcast. Can you give us some idea of the NBC commitment, the scale on which the 1980 Olympics will be done, in terms of personnel and equipment?

DON OHLMEYER: We're going to program 150 hours (approximately twice what was programmed in Montreal) over a period of 16 days. We will probably take in about 600 production and engineering people. There will be a total party of about 1,500 people but that will include sponsors and VIPs and everybody else. We will have more unilateral cameras than American television has had access to in the past. The basic way the Olympics is covered is that there is a world feed which gives your basic coverage of the events for the international audience. Then what happens (which American television really started) is what they call supplementation of the world picture, which means that you come in with your own cameras, and cover more than the host country is covering. By the very nature of having to provide a feed for the world, their coverage will not be good enough for an American audience. Take, as an example, the mile run, and let's say the winner is from Finland. Their basic responsibility is to cover the winner. But let's say there's an American fighting for the bronze medal. The world picture may not pick him up until they get to the finish line, but what American television is concerned with is following that story. You want to show him from the very beginning, even if he's last.

Our normal coverage for a track meet would be about 6 or 7 cameras, and we will have 7 cameras at the Olympic stadium doing what we call supplemental coverage. If we wanted to, we could cover the whole thing ourselves, but the world feed is a great benefit because you don't have to worry about the meat and potatoes in terms of the coverage. You can take care of the gravy, if their coverage is adequate. In Mexico City in 1968, the coverage was really very poor and we at ABC had to do it all ourselves. In Munich, the coverage was excellent, so we were able to spend most of our time supplementing. In Montreal, the coverage was very good, so good in fact that at the swimming pool ABC had only one camera that was their own. The rest was Canadian coverage.

One of the problems we face going into Russia is that there are three different types of television systems in the world. There is PAL, which is used in most of Europe. There is NTSC, which is used in the United States, and there's SECAM, which is used only in France and Russia. The conversion of equipment from one system to another is not that simple, so we're going to take in around 30 cameras of our own. Then we are renting from the Soviets, as has been done in the past, another 30 cameras, all of which will be run by Americans.

We're also taking in videotape equipment, and the interesting thing about the tape is that instead of the 2-inch tape we've used in the past, we'll be using 1-inch tape. That is the state of the art. It's far more flexible and less cumbersome. The problem that you have in television is that the technology outstrips the economics. There is no reason to even use videotape anymore. There are methods of storing an entire show with something that looks like a breadbox, but the problem becomes the economics of changing over. We're just about at the point where your smaller stations have caught up with 2-inch tape, and now a lot of them are being forced to convert for cost reasons to 3/4-inch tape, which is basically not that bad. A lot of the local news that you see is done on 3/4-inch tape.

We'll have 40 tape machines with us in Moscow, and we're building a "master control." When the Olympics are over, the tape machines will be brought back to New York and to Burbank, and will be used to revamp the two television centers with 1-inch machines. The matrix that will be used at the control center in Moscow will be dismantled and brought to Burbank and will become the matrix for Burbank "master control." So most of this equipment that is being purchased now will be brought into service with NBC in the States. Some of the hand-held cameras will go to New York, some will go to Los Angeles, some will be sent to the network owned and operated stations throughout the country to use in their mini-camera operations, and some will probably be sold, depending on how many we can utilize at that point.

QUESTION: What factors do you consider when you determine what you will be programming in your Olympic coverage?

DON OHLMEYER: Well, the main factor is what events take place that day, and we do have a lot of time to fill, so we'll probably be showing a lot more events than have been seen in the past. You also take into consideration the time of day. We're going to have an afternoon show, a late afternoon show, a prime time show, and a late night show. If you take a look at the demographics during those day parts, you find that normally, between 12 and 2 o'clock, 80 per cent of the people watching television are women, but I don't want to do just a woman-oriented show. What I'm hoping is that the 1980 Olympics will be like the World Series when it was on weekdays. People found a way to watch it. It was on in the offices, in the schools, and that's what we want to create for the Olympics. But we will still probably do many more female-oriented features in the afternoon shows.

When we get down to between 4 and 5 in the afternoon, we'll probably do some children-oriented features, but not so children-oriented that it's going to turn off the adults. You walk a very fine line there.

In prime time it's basically what I've been involved in doing for the last three Olympics. We know what works, and there are some other things I've always wanted to try that we will try in prime time. For late night, again you're dealing with a different audience. In late night programming you front load the show because what you are trying to do with a late night show is keep people up. In prime time we will start the show off with something big, and internally promote leading up to something big to go off the air with. In late night, you rank your segments in their order of importance and just air them in that order.

So you do consider the different problems of the different parts of the day, but you have to work within the framework of the events that are taking place on that day. I may feel, "Gee, this would be the perfect time for the 100-meter dash," but the 100-meter dash was run 2 days ago, or it's not going to be run until next Friday. There's really not that much that you can do about that!

SONNY FOX: Let's stay with the Olympics for the moment. What made it so important for NBC to go on the line for an incredible amount of money to secure the rights to the Olympics? They will make it back, as it turns out, but they couldn't be sure of that when they made their bid. They went out and bid $86 million just for the rights to the Games, not to mention the production costs. That's compared to about $25 million that ABC paid for the rights to the Montreal Olympics. To really put it in perspective, the rights for the Olympic Games in Squaw Valley in 1960 cost $50,000. Why was NBC so determined to get the Olympics away from ABC?

DON OHLMEYER: ABC's whole image of superiority in sports has been built around the Olympics. Their slogan at the end of their shows, "Recognized around the world as the leader of sports television," is really an outgrowth of the Olympic Games. They make it sound like somebody voted them the worldwide Emmy award for sports television. It's really something we sat in a room at ABC one day and made up, because everybody else had a slogan. When I first got to NBC we were going to put at the end of some of our shows, "Recognized in Paramus" and let it go at that.

QUESTION: In covering the Olympics in Moscow, how much technology, equipment and knowhow will you have to leave with Russia in order to gain access to the Games?

DON OHLMEYER: First of all, there are no secrets in television. The Russians don't have to come to us to learn the technology of television. An excellent example of that is that Norelco cameras, which are sold in the United States, are made by the same company that makes Philips cameras in France. Philips is the major supplier of cameras to the Soviet Union, so there's nothing we know from a technological standpoint that they don't. That's gobbledygook, and a lot of right-wing propaganda. The same stuff we're taking in they could go to RCA or Ampex or Norelco and buy tomorrow.

In terms of coverage, I think their coverage will probably improve. I'm going to go over and speak to their producers and directors and hopefully share some of my ideas with them. I don't know how the world is going to be any the worse because they may learn how to cover track and field better from American television.

The Other Sports Networks

QUESTION: You've been speaking about the three major networks, but in fact there are other sports networks that provide programming through syndication, like the TVS and Hughes Sports Network. Are those competitors? Are they auxiliaries?

DON OHLMEYER: We don't really consider them competitors. They may consider themselves competitors of ours, but for the most part, unless they develop to the point where they're competing for the major franchises, or for the major numbers in terms of viewers, they are really a satellite type of operation. I'm not taking anything away from them, but when the war councils sit down at the sports departments of the three networks to decide how they're going to bang each other over the head in the next quarter, there is very little thought given to the fact that Mizlou is going to be televising the Peach Bowl or another couple of bowl games, and for the most part they will try to avoid network competition when they can, which is very intelligent on their part. They don't want to go head to head with their golf tournament against a network golf tournament.

It's really more a matter of the network's ability to promote. There's no continuity of viewership for a network that's thrown together on an irregular basis, no continuity of promotion, and promotion is really the key to television. I think that one of the things that Fred Silverman did so well at ABC, as much as designing the programming plans, was that he knew how to promote shows. If you were to sit down and watch promos on each of the three networks, it would be obvious which network is first, which is second, and which is third, just by the promotion.

The Art of Promotion

QUESTION: Don't all of the networks do basically the same kind of promotion for their shows, using newspaper ads and on-air promos and the rest? How does Fred Silverman do it better than someone else?

DON OHLMEYER: It gets down to a question of focusing your efforts, and really thinking ahead. It's doing the Oscar telecast and knowing that Jack Nicholson has a damn good chance to win the Oscar for *One Flew Over the Cuckoo's Nest* and taking the gamble of programming *Five Easy Pieces* for the next night, having a promo that will air immediately following Jack Nicholson winning the Oscar, and being ready, if he loses it, to scrub the promo, or to air a different promo. I'm talking about that kind of thinking.

You can blow a golden opportunity by not promoting it right. As an excellent example, look at the way NBC promoted *The Godfather*. If you go back and look at the print advertisements, it just looks like a rerun of *The Godfather.* In the same typeface that they use to say "Parental discretion advised," they also put, "Footage never before seen." How should you have promoted *The Godfather*? You get Francis Ford Coppola on every talk show saying that "never before in the history of motion pictures has a director's work been presented in its entirety." You could have made *The Godfather* into the television event of the year. Whether people cared about *The Godfather* or not, you could have made the event so important that people would have watched it.

It also ties in with being able to come up with shows that are promotable. There are five sports events that happen each year that everybody knows about. One is the World Series. The others are the Super Bowl, the Rose Bowl, the Kentucky Derby and the Indy 500. So why do you spend a lot of time promoting the Rose Bowl or the Super Bowl? People know it's there. The newspapers are promoting the Super Bowl for you every day in the two weeks before the game, so why waste a lot of time promoting it? Use that time to promote something that needs promotion.

Audiences can be built up through promotion, there's no question about it, but you need to come up with a promo that not only tells people what's on the show, but entices them to watch the show. And that's what Fred Silverman is really good at. My admiration for Fred in terms of promotion really is in the slickness and the quality of his on-air advertising. Before *Roots* aired, there was a two-minute trailer made to promote it that Fred Silverman had the promotion people recut 42 times before he let it air.

Comparing Network Coverage

SONNY FOX: How would you compare the three networks in terms of coverage of sports events?

DON OHLMEYER: I think success breeds success, which is a roundabout way of saying that when I was at ABC I felt that ABC did everything the best. I still think that they do a lot of things the best. I think in terms of coverage, ABC is still number one, NBC is number two, and CBS really doesn't care about coverage. There are certain events CBS is attuned to covering well. The Masters Tournament is probably the finest golf telecast on the air. When they get to the Super Bowl, though, they throw all these cameras in there, but when the director sits down in front of 13 or 14 cameras and 4 slow-motion units and two tape machines and a few other units, he's not going to be able to look at and assimilate the 27 or 28 screens, especially if he's used to assimilating 12 screens. That's why you generally find that in the first half of a Super Bowl the production is really lousy. That's because the director is just trying to get used to the game conditions of looking at 27 or 28 monitors.

I think CBS is just into covering a lot of different events. You see it in their basketball coverage, where they cover a lot of games with just three cameras. They don't really care about the coverage that much. The want to get as many different games into as many different cities as possible so that they can appeal to a wide audience. They show the Lakers game in Los Angeles, the Knicks game in New York, the Bulls game in Chicago, so they've got to cut back on their production costs and they cut back to three cameras. They cover auto racing the same way. They'll go out with three pedestal cameras, or four pedestals and a hand-held, whereas NBC will go out with six pedestals and a hand-held, and ABC might go out with ten pedestals and a hand-held. Those extra cameras help, if you know how to use them.

Bookkeeping

SONNY FOX: As executive producer of NBC sports, do you have to fight with the money people all the time?

DON OHLMEYER: Yes, constantly. Plus the way NBC does its bookkeeping makes it expensive to do shows. According to NBC bookkeeping it costs them as much to do a six-camera pickup in Kansas City as it costs ABC to do a *Monday Night Football* game in Kansas City, with twice as much equipment. It's all in the way they keep the books, and unfortunately we're saddled with it, although it's something we're trying to change. CBS, on the other hand, can go to Kansas City and just farm it out because of their union contract, which neither ABC nor NBC can do. To give you an example, for NBC to do the Long Beach Grand Prix would cost them, from a production standpoint, $450,000 to $500,000 just for the coverage, because they would have to do it with their own engineers. It cost CBS about $135,000. They just farmed it out and some guys came in and laid some cables down and put some cameras up. So even though NBC can rent the trucks and equipment, we've got to put our own personnel in there. And NBC's contract is a very restrictive one.

Capturing the Ambience

SONNY FOX: What are your hopes for NBC sports? You're the executive producer and you're looking down the line at the 1980 Olympics, but besides that, are there any brilliant new ideas percolating?

DON OHLMEYER: It's been very gratifying in the last couple of months that people are starting to notice a basic change in NBC's coverage, especially with NCAA basketball. In the past, NBC's been more attuned to just covering what's happening on the floor, as opposed to capturing the ambience of the event, which is really what makes a sporting event distinctive. There is no question that the Rams play better football, for the most part, than USC, but it's a lot more exciting to go to a USC game than a Rams game, because of the ambience of the stands, because of the cheerleaders, because of the enthusiasm. In college football, the people in the stands really feel that the players out on the field "really represent me," whereas it's very difficult to feel that the Rams represent you. You are going to the Rams game to enjoy the finest in football. You're not going to be totally a part of it, because you can't really identify with those players, whereas in college football, the guys who are out there playing are the same guys you go to lunch with or sit next to in your biology class, or whatever. So if you're televising both college and professional football the same way, then you've missed the boat, because it's two different games, and if you don't capture the ambience, you really haven't captured what you're televising.

SONNY FOX: That ambience can be a problem, it seems to me, in covering something like pro basketball. When you get into one of the big arenas and have to show the fact that half the seats are empty, it has a very deadening effect, and the sense of excitement the announcers try to get into the game in their voices is almost ludicrous when you match it against the reality of the quietude of the stands.

DON OHLMEYER: The game can be just as exciting, but when the viewer sees empty seats, he doesn't really go through a thought process. There's an ingrained sense that says to him, "If nobody goes out to see this in person then why should I bother watching it on television?" And so, I think it loses some of its excitement. The reason people always talk about the Super Bowl being a dull game is because the majority of people in the stadium don't really care who wins or loses, because of the way the tickets are parcelled out. You may have 80,000 seats, and you probably don't have more than 10,000 fans from each team. In fact, after 48 hours of partying, those people couldn't care less what's happening on the field, and it comes across to the viewers.

All of that partying with the sponsors, the clients and the affiliates has become pretty much a part of the sports business. A few years ago the NFL had a party for 5,000 people. Just 5,000 of the most intimate friends, of course.

It's no longer just how well your show will do on the air, but how attractive you can make it to the advertiser to sponsor your program. Part of the appeal of being a major sponsor in golf is being able to go to the locker room and rub elbows with Jack Nicklaus and Arnold Palmer and the rest of the celebrity golfers, and don't think that doesn't play a large part in the minds of the people who control the advertising. You will very rarely find a major advertiser in golf whose chief corporate executive is not a major golf fan, because there's really no way to justify the cost per thousand viewers in buying a golf tournament. They will come up with all kinds of demographics to justify the decision— they'll say that you advertise Cadillacs on golf tournaments because people who play golf buy Cadillacs, but that's a lot of corporate gobbledygook. The bigwigs of the sponsoring corporation get to go to the Masters, and the same thing holds for the Olympics, the Super Bowl or any other pres-tigious sports event.

Responsibility to Minor Sports

QUESTION: Given the tremendous influence of television in building a following for a sport, do you feel that you have a responsibility to cover minor sports?

DON OHLMEYER: I don't think it's just a matter of the net-works deciding what sports they're going to make popular. If it were just a matter of deciding, soccer would be a major sport right now in the United States. It is a matter of giving a sport some exposure, and then the people will decide whether they want to watch it. I feel that women's sports have always gotten short shrift on television. NBC put on the women's collegiate basketball championships from UCLA for the first time in 1978. It wasn't standing on its own as a separate program, but we're planning to do it live in the future and we'll see if it stands on its own. I think we have a responsibility to cover women's sports and to give them exposure. One of the gratifying things about the Olympic Games is that with 150 hours of coverage, you'll be able to see more women's sports than just women's gymnas-tics or women's swimming and diving, which is basically what women's sports coverage has been limited to in the past.

I think we have a responsibility, a responsibility to not only put on what's popular, but to expose sports to the au-dience. I don't think that television can totally on its own create something popular. That is a misconception. There has to be something in it that appeals to the public.

The Future of TV Sports

SONNY FOX: Let's list some of the sports and see what you think of their future growth. Soccer seemed to suddenly take off in 1978, with 72,000 people coming to see the Cos-mos play in New York and so on. How will that affect soccer's coverage by the networks?

DON OHLMEYER: I don't think soccer is consistently drawing big crowds yet, but even if it is, there's a problem. It's the same problem you run into when you're dealing with people in the recording industry. You find a rock star like Peter Frampton, the biggest thing going, who just sold 12,000,000 records, and they can't understand why everybody doesn't want to put his special on in prime time. It's because if everybody who bought his record brings a friend to watch the show, you've still only got about a 12 rating. That's what they don't understand. Even if soccer manages to fill up all the seats in the stadium, it doesn't necessarily mean that it's going to go on television. Television would have a lot more to do with putting people in the ballpark than the people in the ballpark would have to do with putting soccer on television.

I think successful coverage of soccer on television is probably about 7 to 10 years away. The future of soccer is with the interest that is developing among the young kids around the country, and you've got to give these kids, 7-to-10-year-olds, time to grow up. The people who are watching television now, for the most part, have never played soccer. They've never seen soccer played live, and they aren't interested in soccer. It's the whole generation growing up right now who's looking at soccer as an alternative to football.

SONNY FOX: Hockey is even more restrictive in terms of people playing it, because it's a very special sport, and it is usually only popular in the northern climes. Whole areas of the country have never seen a hockey game. Do you see any hope for hockey becoming big?

DON OHLMEYER: No, although it's almost becoming roller derby now, so it could get bigger because of the violence. I'm not a hockey fan, but live hockey games are terrific. It's a thing to be at. The fans are very rabid and, being there, you get caught up in the excitement. It doesn't translate to television well because it's difficult to see the puck, so the camera follows the puck tightly, and that way you lose a lot of the other action. Let's say you're at the game and you see a guy skating across the ice from point A to point B, and you know he is going to slam into the guy who is positioned at point B. Now, just before he gets there, the guy at point B passes the puck. The camera, having no peripheral vision, follows the puck, and maybe just as the guys are going out of frame, you'd see this guy going up into the stands. If you were at the game you would watch the hit and your peripheral vision would be following the puck. That's the same problem that comes up with baseball on television. There's no peripheral vision.

SONNY FOX: What about tennis? That remains a sport with a very limited appeal.

DON OHLMEYER: Tennis has two problems on television. One is that generally when it's on the people who would watch it are out doing it. The other is that there was a fanatical burst of enthusiasm for tennis about five years ago, and the number of players went from 5 million to 30 million. I just read an article in the paper which said that over the last 2 years, they lost about 10 million tennis players. A lot of people went out and bought shoes and a shirt and pants and a tennis racket, went out to play and said, "Wait a minute, this isn't all that simple. I'm not coming out here every Saturday morning and making a fool of myself! I'll go back to making a fool of myself at a game I like . . . golf." Tennis does not necessarily have the universal appeal of other sports. In addition, it's hard to cover it well on television. You can sit and watch a tennis match on television and have a very difficult time seeing the ball.

SONNY FOX: So far you've basically shot down soccer, hockey and tennis. Are there any other sports that you see getting more coverage on television in the next 5 or 10 years?

DON OHLMEYER: I think the next big burst is going to be in women's sports. I'm not sure which women's sport it will be, but there will be a major increase in interest in women's sports, if only because women have come out of the closet about participating in sports. A woman is no longer considered "butch" if she plays sports whereas, 10 or 15 years ago, if you were a woman basketball player they thought you were strange. I think there are a lot of women out there who have started to enjoy watching women's sports, and I think men enjoy watching as well. I would much rather see a tennis match between Chris Evert and Billie Jean King than I would between Jimmy Connors and probably anybody else. I play tennis, but the game I play is not the game Jimmy Connors plays. The game I play somewhat resembles the game Chris Evert and Billy Jean King play.

Sports Salaries and TV

QUESTION: Would you say that there's a strong relationship between television ratings and athletes' salaries?

DON OHLMEYER: Yes, and I think this is ultimately going to be why pay television will be the only answer. What happens generally is that the networks sign 4-year contracts with the sports franchises. In the first 2 years of the contracts, the owners make out tremendously. It takes the players about 3 years to catch up, but when they do, they get these enormous raises you read so much about. I would think that by the end of the most recent NFL contract you will probably see the salaries of professional football players doubled. So far the networks have accepted the burden of paying more and more on the rights deals to pay those salaries and to keep the owners happy. But those salaries will increase.

Basketball now has a tremendous problem. CBS is paying a large sum of money for the NBA contract, and it's not doing very well. Consequently they can't sell the time on a basketball game for as much as they would like to, and when the next contract comes up, that huge increase in the rights fee may not be there for basketball. So the only way the team owners can raise revenue substantially is by raising ticket prices, and ticket prices for sports right now have priced the fan out of the marketplace already. The individual fan just doesn't make enough money to go to very many games, and if we ever get into a tight economy where the corporations start to cut back on their season ticket purchases, a lot of teams are in trouble. Sports are so intertwined right now with television revenue that if the networks can't continue to escalate the prices they pay, the sports owners are caught in a very precarious balance. Pay television is their only way out.

News

Richard C. Wald

Richard Wald has enjoyed a distinguished career in both print and broadcast journalism. A graduate of the Columbia School of Journalism, he first became affiliated with the *New York Herald Tribune* as Columbia College correspondent in 1951, and continued to serve the newspaper until 1966, successively holding the positions of religion editor, political reporter, foreign correspondent, associate editor and managing editor.

Wald was Executive Vice President of Whitney Communications, Inc. 1967-68, and joined NBC News in 1968 as Vice President of the news division. He became President of NBC News in 1973, a position which he held until 1977. After a brief stint with the Times-Mirror Company in Los Angeles, during which time he made the remarks which follow, he returned to New York and television news as Senior Vice President for News at ABC.

Television is essentially a process in which we gather a very large crowd of people in their own homes to listen to what an advertiser has to tell them. In between the times when the advertiser is talking to them, we have to do something to keep their attention. In general, that will be entertainment programming—the essential, central avenue of television. For complicated reasons (none of them the ones publicly stated) the news program sprang up. The people who run television companies have always thought that TV had inherent in it much more possibility than the "least objectionable programming" that Paul Klein speaks about, and the owners and main managers thought that news was of value.

News is always peripheral to entertainment. Television is not concerned with how much money you've spent on programming so much as how much time it gets on the air. The essential criterion, the non-elastic substance of television, is time. There are only 24 hours in a day. There are within that day 2 1/2– 3 hours of "prime time" as we define it, and say 6 or 8 really prime hours of broadcasting per day. In those hours what gets on the air is what is most prized, and you can be sure that news gets on less than entertainment. Nonetheless, for the time it does get on, news occupies a very large role in the three networks. If you run a local network affiliate station, it is entirely possible that the majority of the local programming you are airing is news. Independent stations will program almost everything themselves, but a network affiliate, getting the network feed daytime, early morning and prime time, will have a lesser amount of the day to fill, and that station's image will be strongly influenced and characterized by its news product.

In all three networks, the television news operation is closely allied to the radio news operation. In terms of the number of men and women working for the network, those divisions tend to be very large. You've got to remember that networks buy nearly all of their entertainment product from the outside, so their main employees are technical employees to keep the network facility going, various kinds of service employees, executives in various shapes and sizes, and news employees. A news division for any of the three networks will run somewhere around 750 people, meaning mainly people directly involved with the news (not counting people who maintain the electronic interconnection), and those people are spread around the world. There are, on the average, about 7-10 bureaus in the U.S. for each network, and say another 12-14 of various sizes overseas. They will spend, depending on the year, between 70 and 100 million dollars. The upper limit is reached in a presidential election year, because conventions and election nights are extremely expensive to put on. And they will employ, in terms of the time given to them, a half hour a night, usually seven nights a week—that's 3 1/2 hours a week. All three networks have an early morning news program. CBS only has an hour in the morning, so that's 5 additional hours a week. NBC does 2 hours in the morning, so that's an additional 10 hours a week for them, and ABC has the *Good Morning America* program, which also contains substantial news coverage, even though it originates in the entertainment department. Each network does roughly a minimum of 12 hours of documentary programming a year, and CBS does the most with about 24 hours. At the moment, these things are peripheral to the main business of television.

The Rise of TV News

One of the interesting things to remember about television is that it is one of the most widespread institutions in our lives which is entirely a product of technology. There would be no television if someone had not invented all the technology that makes television, and it's relatively new technology. Although it was possible to create television signals before World War II, the real growth of television as we know it today happened after that, from 1945 on. Network news as we understand it in television terms really began with the 1948 Democratic convention in Philadelphia. They decided in '48 to allow television cameras into that convention, because the Democrats realized that the coaxial cable passed through Philadelphia, and you were able to get at that time through the entire U.S. somewhere between 3 and 4 million television homes. In order to broadcast from the convention, the networks all needed television people and news people, and that's when they really began staffing up. The 1952 conventions were better planned, required more staff, and were really responsible for a lot of pressure on these news organizations to organize themselves intelligently and to create themselves as news-gathering—news-reporting operations. If you figure that 1948 was the real beginning, that's about 30 years ago. It has never happened before that an institution only 30 years old bulked that large in a nation's consciousness. And yet, the technology that gave birth to this thing is right now in the process of an explosive amount of change.

New Technology and Networks

Consider briefly that the business of the network is to gather an audience and give it to an advertiser. There are a lot of things changing the networks. There were about a million cassette machines in civilian hands at the end of 1978, and a dangerous weapon they will be, because you will not need very much effort to tape a program and play it later, or to tape a program while you're away, and play it at some other time. It will take a bite out of the mass of the audience at the time a program is broadcast. We are moving toward the introduction of disc machines into the market which will produce what look like LP records which will cost

$7.50 each and will play for an hour on your television screen. The number of home television cameras being sold is going up, in some ways comparable to the number of home movie cameras sold when they were first introduced. They're very expensive—it's a rich man's toy—but it's the sort of thing you can see becoming a little, or maybe even a lot, more accessible to a mass audience. The number of television games sold is enormous. The number of households on cable TV will reach 1/3 of the U.S. by 1981. Roughly 70,000,000 people will have a cable, and on cable you will get 3 dozen channels instead of 4 or 5 or 6, and that will fractionate the amount of attention given to any given network. Much of the cable programming will simply reproduce network material, but a lot of cable TV will also be devoted to peripheral information items—time check, stock market reports, whatever other additional stuff is available.

There are also experiments going on with different kinds of television receivers, so that it may be possible in the fairly near future to split the spectrum. That would enable us to double the number of VHF stations and to double the number of UHF stations. Certainly it will be possible to receive UHF stations with the same clarity and ease that you receive VHF stations, in which case the number of choices available to an audience will expand greatly. In addition to all of that, there is that never-neverland kind of stuff that they're beginning to talk about in terms of home computers. You can now buy enough computer material to plug in to your television set and have it do your income tax. It's expensive, but you can do it, and that technology will get less expensive and more accessible.

Each one of these things takes a bite out of the mass use of the television screen for the networks. It isn't going to destroy the networks. They're going to have a healthy operation in the future, but it's going to change the perspective with which networks are seen. It will provide a great deal more in the way of competition. The effect of that on both entertainment and news is really a little cloudy. I believe that the networks will remain the dominant force in terms of television entertainment because they have the capital and a lot of people with experience in dealing with mass entertainment. One of the things, however, that you can't put in a tape or in a can, that you can't bicycle around to station after station, that you can't do without a lot of people involved, steadily employed all the time on staff, is news. As the networks change their balance, they're going to have to find new ways to put entertainment before the public, and new ways to deal with entertainment. They're going to have to put a different kind of emphasis on what they do because as cable comes up, pay television will come up, and as pay television comes up, it may take over some of the aspects of present free broadcasting. You can easily see a system in which a lot of sports broadcasting goes to pay television. As these balances change, the one thing you can't change too much is the basic news operation. If you want that kind of news, the only way you can do it is to have the bureaus in the U.S., the bureaus overseas, the light-weight cameras, the delivery systems and the editing systems that are presently in place in the networks. What I think that will mean is that the news aspect of network life will get larger, because it will become, almost in parallel to what happened in radio, the central point in a much more diverse universe. In radio today the situation is that news is the central part of network operation, whereas 30 years ago news was peripheral to the entertainment programming of radio. It isn't the central part of all radio, but insofar as networks exist in radio, they're mainly news networks. Now something similar will happen in television. As the multiplicity of outlets grows, the central core of what the television network does will become more and more news, because that's the main thing that you do that you can't prepare in advance and package.

I don't know what that means to audiences either. As you split an audience among a multiplicity of possible outlets, you also reduce the number who will be watching any given channel. As that happens, the amount of money you can get back from an advertiser goes down, and as you spend less on those things, how much are you going to spend on news? I don't know whether these very large news staffs will continue into the future. I suspect that they will continue longer than the equivalent very large entertainment centers do, but I'm not positive. I would guess offhand that the news operations will become somewhat decentralized, although they will never go into the kind of independent programming that entertainment will be.

How Much Time and Money for News?

SONNY FOX: Do networks make or lose money on news now?

RICHARD WALD: In the main, networks lose money on news. That means in any 4-year cycle they lose money, because every 4 years contains within it a national election year, and national election years are big loss leaders. In any given year inside the cycle, they may make some money. For instance, in 1978 it is entirely possible that the news operation at NBC or CBS, and maybe even ABC, could make money. CBS probably does because it has a highly successful news magazine in prime time, *60 Minutes*, it has a fairly successful morning program, and it has a very successful half-hour news program in the Cronkite program. So that all together, in a relatively stable year, they probably make some money.

This figuring includes the cost of documentaries, but it doesn't include some worldwide event that begins to cost you an enormous amount of money. If an event like the Sadat visit to Israel and the whole process takes two weeks, it's affordable. If a thing like that were to take two months, you would lose everything you could possibly make in profits for the year.

I'll give you an instance of how money gets spent. When former President Eisenhower was in Walter Reade Hospital for the final months of his illness, it began as a network practice for all three networks to have portable electronic equipment outside the hospital door. Very, very quickly they formed various arrangements and pools, because within a couple of days each network was spending at the rate of a couple of hundred thousand dollars a week just on that coverage. If you have one of these very large organizations you can do it, but you can't do it for very long without spending one hell of a lot of money. When we did network coverage of the Watergate hearings, it not only cost money to mount the coverage of the hearing, but the networks also gave up all of the income they would have made from the programming that would have been on the air during the period the unsponsored hearings were on. The cost then was not only the pool share to each network, but the additional cost of not doing any commercial business during those hours. That cost can mount up.

SONNY FOX: That kind of expenditure represents a commitment then by the networks in terms of public service, which leads me to another critical question. When something comes up that you feel, as head of the news department, ought to be covered and needs to preempt other programming, at what level is that decision made, and what goes into the decision-making process?

RICHARD WALD: This is a question that can come up fairly often depending on the news weather. If it's a stormy news year, it comes up very often, but it comes up in imprecise ways. If you get a situation where the President of the United States is resigning, nobody says to you, "Is that worth covering?" But if there's a new development in the Watergate situation, there is a very fine line as to whether or not that is worth special coverage. Say the Supreme Court says Nixon has to produce the tapes. Well, is that worth a special or isn't it? You make a judgment based essentially on a news sense. You cover, for instance, presidential press conferences. Do you cover a press conference when the Secretary of State speaks? Sometimes you do, sometimes you don't. It depends on what it sounds like.

Let's assume that you decide that this event is worth covering. You then go to the president of the company. The way all three networks are set up (except at CBS where it's slightly murkier) the head of the news division reports directly to the chief executive officer of the television company. He goes to the guy at the top and says, "Look, we need to do a special on the fact that the Supreme Court says that Nixon must produce his tapes for the trial." If the president agrees with you right away, he then says okay. Remember, in each instance, even if they give you the worst time available, a half hour at midnight, it's going to cost them a minimum of $60,000. If you can persuade them to give you an hour in prime time, it might cost $400,000. So you're really playing with very big chips. In the case of a wild catastrophe—war is declared, peace is announced, there's been an atomic explosion, the President is resigning, almost anything like that—you don't even ask. You just take the air, and you go. We didn't ask, for instance when former President Johnson died. We just interrupted normal programming and went on the air. We didn't ask anybody. We didn't have to. But in those things that are marginal—not marginal in terms of news, because you know it's news, but marginal in terms of being a special event—you go to your president.

He may say he's not convinced. He thinks maybe you're right, but wonders what other considerations there are, so he then calls in the president of the television network, who is the man in charge of the entertainment programming, and he says, "Joe, we have a question here. News wants to do a special, and they want to do it in prime time this evening." And Joe says, "Oh, my God." He always says, "Oh, my God." Institutionally, the guy who runs the news is always trying to get news on the air, and the guy who runs entertainment is always trying to get entertainment on the air. They may love each other, and outside the office they may share an identical view of the world, but in the office they fight, and what they fight about is a very subtle series of questions.

If you look at it over any 4-year period, news loses money. Whose money does it lose? It loses the entertainment division's money, the money made by that fellow who runs the network. There's another thing to consider. Not everybody in the country is enamored of news. All right-thinking people are, but not everybody is enamored of news. If you interrupt a program, a percentage of the audience always resents it. Always, on any program. Now, if you interrupt programming to say that the President has just fired the Attorney General, people will say, "Gee, thanks for telling me," in the main, but a sizable proportion of them will say,

"What else is new? Why did you take away the movie?" Or, "I spend my whole week looking forward to *M*A*S*H* and they interrupted it." Or, "Why is it the President always has press conferences during *As the World Turns*?" And that's a real question. The people are entitled to have essentially what they've been told they're going to have. Why shouldn't they? It isn't written anywhere that they have to watch a press conference. That's not part of your citizenship papers.

Then you get to another consideration in the business of the network which is kind of interesting. The president of the network will say, "Look, boss, you told me that in order for all of us to be happy and successful I have to win a ratings battle. The ratings battle is judged cumulatively for the week, and we have three hot one-hour shows on tonight. Each one of them is going to get better than our average rating. If you let this news guy take one of those hours, then our ratings are going to go right through the ground, because not only will he ruin the hour he's taking—people don't want to watch news—but if we put it on early, the audience won't be there for this other terrific show we just paid $500,000 for. And if we put it on late we'll lose the audience for all the news programs for all the affiliates. It's terrible. You can't do it." Then the boss says, "Well, how much will it cost if we do it?" And from his back pocket Joe, who knew this question was coming up anyway, pulls out a sheet and rattles off how much it will cost. Then you begin the argument. What you do essentially is argue it out on the basis of the various merits concerned. Everybody has prejudices. At different times in different places for different reasons, chief executive officers lean more toward news, or lean away from news. Or news people are more persuasive or less persuasive. A lot of it comes to that kind of human equation. A lot of it also comes to a question of: "For God's sake, we haven't had a special on in 6 months, so we can do it," or, "For God's sake, we had six specials on last week, so we can't do it."

The question of what the competition is doing is also a main point. If you think the competition is going to cover the story or not, either way, you're in a double bind. If they're going to do it, you really want to do it, because you don't want to be beaten by them, and that's a persuasive argument. On the other hand, in the abstract, if the competition is going to do it, the public will be informed, so you don't *have* to do it. If you don't think the competition's going to do it at all, you have the opposite possibility. You're going to beat them. At long last, you're really going to slip it to them. But, on the other hand, if they're not doing it, why should we? And if they're not doing it, they'll clobber our program in the ratings.

It's a finely graded series of interesting questions, all of them argued out in the space of an hour, or a half hour, because you don't have much time. Actually, winning and losing get all mixed up because losing fights sometimes wins things for you later on.

One of the major problems for all three networks is the question of the time slot at 11:30 at night. You get into a situation where that peripheral news program finds an easy resting place at 11:30, because in a very tight race the networks will attempt as much as possible to resist news on a special basis in the earlier hours of prime time programming. That creates another kind of difficulty with the network that has the best results at night, namely NBC, which has a very valuable, much-watched program in the Johnny Carson show. Every time they delay the Johnny Carson show, they risk driving the ratings down.

The 11:30 Slot and "Weekend"

Putting your news special on at 11:30 has an interesting consequence. You lose a large part of the audience. The number of people who are up at that hour who want to watch a news program is, of course, much smaller than the number of people who are up an hour earlier. You begin to trade off a whole lot of things, and some stories are just not worth putting on at 11:30 at night, even though they might be worth putting on at 8 o'clock. That becomes a funny value question that faces television people, and never print people, because a newspaper's a newspaper's a newspaper. It's always the same paper. The time of day when somebody buys or reads the paper is any time when he or she can buy and read the paper. But television, because of the inexorable quantum of time, is different at different parts of the day, and television news is subtly changed by the time period in which it appears.

That accounts for another interesting property of news that most people don't realize. About five years ago, while I was president of NBC News, NBC decided to explore the possibility of having a news magazine at 11:30 on Saturday night. The available audience then was almost nil, and it was my judgment, backed up by Reuven Frank, who was my predecessor as president of NBC News and subsequently became a television producer, that you could *create* an audience. Nobody had ever created one for that time period. It was the network's judgment, after a lot of arguing back and forth, that it would be cheaper to let me make a fool of myself and give me the money to produce a news magazine program in that time period, because then I wouldn't have the money to produce specials and documentaries to interrupt their prime-time programming and cut their ratings.

So we made a trade-off and for the money I had, which was actually a lot of money, I bought 12 programs, 90 minutes long, to air once a month at 11:30 on Saturday nights. At the end of the year, there was enough of an audience there for the network to realize that instead of giving me all four Saturday nights per month, which is what I wanted, it was too valuable a time slot to give to a dummy like me. Thus *Saturday Night Live* was born, because they saw the possibility of creating an audience that would be younger, hipper and would stay up late.

The news program we put into that time slot, called *Weekend*, was different from the normal news programs. It was unheard of in news programming to have an anchorman who didn't wear a tie, and we made a point of this as an opening advertisement to show that this was not like other news programs. *Weekend* was done in odd ways, and took up odd subjects. Because of various financial constraints, individual segments were a little longer than I would have wanted them, but that really doesn't matter. Essentially, it was an attempt to create an alternative kind of news programming. One of the great experiments at NBC in the

1978-1979 season was taking that *Weekend* program out of the nice comfortable Saturday night ghetto it inhabited and putting it into prime time. If they take the same program and put it into prime time, will it fly? And if they change the program, will it be any good? And if they change the program, how should they change it? These are very interesting and complicated questions because there are various possible answers.

We tried twice to get that program on the air in an early evening time period before, but we ended up in a battle with the local stations, the affiliates. This happens frequently when the networks try to take over time which has been programmed on the local level. When we lost out on the early evening slot, Saturday night was my consolation prize. We had, more or less, worked out within the network that we would try to take Saturday afternoon from 6 to 7 for network news magazine programming. That is not prime time—it's before the local stations' Saturday news programming. Local stations have a finite amount of time which they normally program. Theoretically they could program the entire 24-hour day if they wanted to preempt their network contract programs, yet an essential politeness and a bit of greed keep them inside the network fold. Essentially they will run the network offering, except for the truly disastrous shows which, after failing locally, they will often replace with local programs. Within a schedule basically comprised of network programming, the station makes more money running any given local programming than they make from a network show, because they get higher rates from local advertisers, they can sell all the minutes in the time period, and they can produce programs inexpensively.

Local station prime time tends to run from 6 to 8 o'clock, because it comes just before network prime time. Local stations make a lot of their money there. In that 2-hour period it is not unusual for a station to make 50 percent of its revenue. Even though the stations are paid by the network for running network programming, in the form of network compensation, our news magazine would be eating up valuable local time. There are basically these local station arguments against doing it.

First, they say, "We have a regularly scheduled news program time. If you take that time you kill our news program. Killing our news program for your news program is a bad idea. We offer a local service, and therefore you can't do it."

That a perfectly reasonable, respectable argument, the one normally made by most stations. What most of the stations mean, however, is, "I'll make more money if I program locally in that time period, and I am not going to give it to you. I don't have to, and I won't."

There is another argument, and it is real and true, and that argument is, "Look, too much of our air time, news or whatever, is done by the networks, and we have to live in the local situation. And we have to have a sense that this is our station. That's all—it's just our station, and you can't have any more time, because the more time you have the less it is a station of ours."

All those things together meant that there would not have been enough clearance on Saturday afternoon from 6 to 7, and so we got shot down.

We then instituted the *Weekend* program at 11:30 at night. In order to make it fly with as many station clearances as we could, because that too was station time, we allowed stations to program it either Saturday night or Sunday night. Hence the name *Weekend*, because you couldn't call it *Saturday Night* and you couldn't call it *Sunday Night*. A couple of years after we started it, and it became successful, *Saturday Night Live* started up. The network wanted that additional Saturday night. At that time I was willing to give it to them if I could get additional time elsewhere. We began to talk about the possibility of doing a news magazine program late Sunday afternoon. We ran into the identical problem we had in the Saturday evening fight, but even more virulent, because on Sundays the networks had taken to programming not only baseball and football games, but tennis, golf, soccer, ping-pong, anything you could think of, because sports in that time slot is a relatively good draw and it makes a reasonable amount of money, and the networks were very interested in promoting it. The local stations, however, were losing their own time to the network sports programming. They didn't mind that too much, as long as the audiences were very good, but they weren't interested in losing even more time with our news magazine. And, in addition to losing that time, they initially at least would not get a good rating. So that, all in all, we never got the time. The stations had the clout to keep the networks from getting the time.

The One-Hour Nightly News

When the question of a one-hour nightly news program comes up, the problem becomes even more apparent. All three networks tried at the same time to develop an extended nightly news program. CBS went so far as to produce a dummy program. NBC went so far as to outline a dummy program. ABC went so far as to outline its own program, but for 45 minutes, not an hour. All, however, were thinking of a longer time period. Stations came in and made their arguments, and they made quite persuasive arguments.

"You're taking away from our news time. You're taking away from the revenue that helps us to exist. You're taking away from local programming, whether or not it's news time. You are not providing us with any great alternative service, just more of the same. You haven't invented anything brand new."

Each one of those arguments carries a certain amount of weight. The greatest amount of weight, however, is, "No, we won't do it," and they would not, so the networks were forced to abandon their plans. I told the NBC affiliates at that time, in 1976, that at some point not too far off—probably within 5-7 years—I believe the stations will come to the networks and ask for an hour of news. They'll do it because news will be one of the few things that will not be threatened by alternative television delivery systems in their areas. It will be something unique, in the face of the technological changes I was talking about which will spread out the marketplace and require a great deal more ingenuity in programming. But until that time when they come asking, *if* they come asking, the networks will not have a one-hour news program.

Television News and the Newspapers

QUESTION: How do you think the growth of television news has affected the newspaper business?

RICHARD WALD: As you know, television has become tremendously important in the news sense. More than half the country says it gets its news from television. Clearly the people trust television, clearly they're getting their news from television. What kind of news, how much news, and on what basis is very hard to say. They get better news from television than most people on newspapers like to think. They get pretty good news from television, but for a period of time newspapers were uneasy about what they would do to combat television. They used to have a monopoly on the mass distribution of news. Then television came along and took a big bite out of their market. Newspapers floundered a lot, and did a lot of odd things in response.

I think generally they have decided that the worst effect of television news—or maybe the best—has been on afternoon newspapers. If the newspaper comes into your home at 4 or 5 o'clock, it had to be printed at least an hour earlier. If it was printed at 3 o'clock, and local television news is on at 6 o'clock, television has at least 3 more hours update from the wire services, reporters' input and everything else, and it can do a better job of giving you the headlines than the afternoon paper can. The newspapers decided that they'd better not try to compete in headlines so much as in specialties. They went after great writers, great thinkers, comics, all that sort of stuff, and those afternoon papers that have done well by and large have done well with formulas like that. Some of them have also done well, though, by offering a detailed kind of news that television cannot.

Now I think we should get into an area which is crucial to the way news operates, as differentiated from television, newspapers or magazines as particular media. Try to think for a minute about the fact that news is independent of the medium through which it is transmitted. Just imagine, if you can, that it is shapeless, it is gossip, it is what one person would tell another if they met in the hallway. "Gertie is pregnant" could be the news item for today, and the way a newspaper handles it is different from the way television or a magazine handles it. The manner of presentation is different from the matter of presentation. You don't have to be able to read to watch the news on television or hear it on radio. There is less work and less learned ability called for to comprehend information presented by the broadcast media.

The manner of presentation also is totally linear. When you say, "Gertie is pregnant," one word comes out after the other and once you've said it, it's gone. It's a little like music. You have to remember what the themes were in a symphony because no one is going to go back and play it again. In addition to that, of course, in television we have pictures. It is very visual, but it is not pointedly detailed. If I show you a picture of a pregnant woman, you will see that she is pregnant, and depending on who you are, you will see what she's wearing, or how her hair is done, or what her background is, or whether she looks good or bad—whatever catches your eye. But television doesn't tell you what to see.

In print, however, it is non-linear. If you want to go back and see something more clearly, you just read the paper again. It is also non-time-specific. You can carry the printed thing around with you and get to Gertie's pregnancy later on. And it is directed, in the sense that what I tell you about it is all you know. If I tell you that she's pregnant and is wearing a blue dress, that's all you know about it.

There are, I believe (though I'm not really sure of this), people who prefer their information in one way as against the other way. There is always a minimal segment of the population who just can't take it in from television and want to take it in from print, or vice versa. There are also subjects that are particularly appropriate to one medium or the other. It is very hard to transmit detailed financial information in television, but in newspapers it's easy. You can print all the tables and the graphs, and if you don't want to see it you can skip it, and if you do want to see it you can go over it again and again. Similarly, it is very hard to do justice to a moon landing in print, whereas on television you can share the experience. The more experiential a news item is, the better suited it is to television. Maybe not the better done, but the better suited. The more abstract it is, the more non-immediate in feeling it is, the better suited it is to print. You've got to remember that print, no matter how well you think of it, is very abstract. It's just a bunch of black lines against a white background. You have to learn how to interpret that stuff. You've got to remember that the word "apple" means a thing—but it's either yellow or green or red. It's an abstract of the thing, whereas a picture tends to be looked on as the thing itself. A picture is very concrete. I think there's going to be a general separation, as there is now, of those things suitable to television, which will get most television time, and those things suitable to print, which will get most print space. And there will be a gray area in the middle, which will depend on how well each medium covers it.

Choosing an Anchor Person

SONNY FOX: In 1976 Barbara Walters left NBC, where she was doing the *Today Show*, and went to ABC and made her famous million dollars a year to become co-anchor in the 7 o'clock nighttime news show. Did the issue arise at that time at NBC about making her an anchor person there?

RICHARD WALD: We did discuss that. It was a complicated process, but the simplest answer is that we had to look at where we were. We had a successful prime time anchor team and we needed Barbara to do the *Today* program, not to do the nighttime news. ABC had an unsuccessful nighttime news program and wasn't really interested in daytime. So they wanted Barbara to improve their nighttime programming. Maybe it would work, and maybe it wouldn't. If it worked, they would be helped in the nighttime, and at the same time their daytime morning program, just starting, would have an advantage, because ABC would have seduced away one of NBC's major stars. So they would have gotten a double advantage out of it. If she didn't do well in the nighttime, they would still have given their morning program a good head start against the entrenched morning program at NBC. So it was a good strategy for them to follow.

SONNY FOX: Where you flabbergasted at the money?

RICHARD WALD: Well, sort of, but there's a great line in a Yeats poem which says, "All things tend upward," and I think that's true here.

SONNY FOX: Dick Salant, your opposite number at CBS, was outraged, and thought it was a crime against the news that someone would make stars out of these people and pay them a million dollars.

RICHARD WALD: Walter Cronkite was outraged as well. So were some of the ABC people, who weren't getting a million dollars. But you know, upon quiet reflection, almost all of them took the same general view. If Barbara Walters is in news, and she makes a million dollars, there's a possibility I could do it too. And once that dawned on them, they all began to think of her as Joan of Arc, giving her body for their good.

SONNY FOX: I think what really had Dick Salant concerned was, if Barbara Walters was worth a million dollars, in heaven's name what was Walter Cronkite worth? Harry Reasoner did, in fact, get a very substantial raise—they doubled his salary. Now, having done all that, it didn't work. Why?

RICHARD WALD: Well, there are a lot of reasons why. It is difficult to pull it apart scientifically. There are different ingredients in a successful news program. First you have to have a good news program. That may sound simple, but you really have to have a good organization that produces the news interestingly, timely, competently. I don't think ABC had the intra-structure, when Barbara started, to produce a good news program, at the level that NBC and CBS were doing it. Then, you have to have a popular anchor person. Everybody was buzzing recently because Roone Arledge, the president of ABC News, decided that ABC was not going to have anchor people of the traditional nature. Instead they're going to have people who, as Howard K. Smith said, are like characters in a Punch and Judy show. They pop up, they say a few words, and then they go away. Roone will un-

doubtedly have an anchor person as soon as he finds someone he wants as an anchor. Until then, he'll hold this new
philosophy, but when he finds the person, he'll change the
philosophy. It's perfectly reasonable to do it that way. He has
a problem, and this is the only way he can solve it at the
moment. Very pragmatic, as everything is in successful
television.

You have to have a good anchor person, but anchor people
are seen wrongly, I think. Anchor people are to television
news programs roughly what layout and makeup are to
magazines and newspapers. They are the visible embodi-
ment of what you're doing. If you have a screaming headline
and wild makeup on a newspaper, but everything is written
as though it's being done for the *Morning Telegraph* of
1890, you won't have a successful match, and it won't succeed. If you have a very sedate layout and it looks like the
New York Times of 20 years ago, but it's filled with nothing
but sex and violence, you really won't have a successful
match, because people who like sex and violence don't particularly like that layout. Anchor people serve as an indi-
cator of the type of news presentation the viewer can expect,
and in addition they act like shunt engines in a freight yard.
They move the stuff back and forth—they take it from here
to there. They serve a dual function.

The match between Barbara Walters and Harry Reasoner
did not seem to work out. The pairing didn't work and the
news product of ABC also didn't work, so they had a series
of difficulties, and after a while it became self-defeating. You
get a sort of mystical feeling that it isn't working, it doesn't
work, but you don't know why. I am widely quoted, prophet-
ically, as saying it wouldn't work. I didn't think it would
work for various reasons. One of them was that I felt that
Barbara's strength was as an interviewer, and not as a person to sit there and be a shunt engine. It's not that she
couldn't do it, or do it as well as a whole lot of other people.
It's just that her strength is imprinted on an audience's
mind—interviewing is what they've come to see. If she does
something else, they say, "Skip that, let's get to the main
act." I just didn't think it would be good enough—she would
not be outstanding. She wouldn't be five times better than

Walter Cronkite, and short of that she would not be living up to the public's expectations. God knows, she's an attractive on-the-air personality, and a terrific interviewer, and she is very, very good at what she does. But she is not particularly good as an anchor person, a very boring job. Everybody who does it says the same thing, and says it truthfully, that it's a job in which you were raised to be a race car and all of a sudden there you are as a shunt engine, pushing things around. Barbara is a woman of great interests and intelligence and a great deal of energy, and I didn't think she'd be happy as an anchor person even if she was successful.

Walter Cronkite finds a great deal of satisfaction despite the problems and the boredom of the job in being, in fact, the managing editor of that program. He has a hand in shaping it, and that shaping hand is of value to him—he takes pleasure in that. In addition, the match between him and the program is almost perfect in terms of CBS requirements. The kind of news they present is Walter Cronkite kind of news, and they have Walter Cronkite there to do it for them. Insofar as they keep the match in balance, and maintain their very good staff and their very good news program, they do very well. It isn't Walter Cronkite all by himself, although that's the way it looks. You look at Mt. Everest and you say, "Gee whiz, that's very high!" but it's that whole big mountain range that's impressive. Well, it's the same thing with CBS and NBC. In NBC's most successful days, the Huntley-Brinkley days, the kind of news NBC presented was a slightly irreverent, occasionally serious, always lively news program. That described the two people who were the anchors, and the news itself was tailored that way. When they parted, the news continued the way it had been done before, and different anchors came on, but the match wasn't as perfect. And when the match stopped being as perfect, ratings began to go down. They never went down very far, but you always have to remember that there's only about one rating point, sometimes two rating points, between the leader and number two, and then there are five to seven rating points down to number three. The difference is minuscule. A difference in a rating point is about 700,000 people nationally, out of a total audience of approximately 12,000,000 people.

SONNY FOX: How do you go about finding the right person? Do you change the news to match the person, or do you try to find the person to match the news that you now have? Who do you put in there? When John Chancellor leaves his spot as anchor of NBC's evening news, how do you go about replacing him?

RICHARD WALD: Well, you don't change the news as the first step. You look at the organization that you have and you say, "What are we? Where are we going? What are we going to be two years from now?" and you come to a series of conclusions. NBC decided that it would like to be both a purveyor of headlines, of the excitement and the hard news, and a news program that will make an attempt to get behind the headlines at least once a night in at least some stories. NBC decided to try to provide the audience with a sense that something from the show sticks in the mind, that it is not just a headline service, it really gets at something every single night, except occasionally when a big story comes along and the whole program deals with the big story.

If you stay that way, and I presume NBC will stay that way—it's a brilliant idea—you have to look to what kind of anchoring will provide the best match for that. Because John Chancellor wants to put on his trenchcoat again, that does not mean he was a bad anchor, it only means he wants out. You then begin to look for someone who will suit the way you want to go, and consider how Mr. X as an anchorman will change your direction.

Suppose Barbara Walters was still at NBC when this question arose, and was doing the *Today* program, and you said to yourself, "Should we have Barbara Walters as an anchor?" I think the situation at NBC is such that she might make a good anchor, because she would not have to do the shunt engine bit she was doing at ABC. NBC has more flexibility. It isn't building an organization, it's got one. It might well be, if you wanted to structure the program slightly differently, that you could take Barbara Walters and say, "Look, your strength is doing those very penetrating, tough-minded, sharp things you do. Suppose we decide that the way we want to do a 'Segment 3' kind of piece is not the way we're doing it now, as a sort of five-minute-long mini-documentary per night, or three nights in a row, but we want to give the world an insight into a particular problem each time, and you're in charge of getting it done for us."

Five minutes on a 24-minute program is one heck of a lot of time! It can be the centerpiece of a major operation. It might be very interesting to have Walters doing that thing, and somebody else, somebody good like a David Brinkley, doing the rest of it. Or you might say to yourself, "Look, we want to do a program with more background in it, but to make that stand out as much as possible, we want to have a program that deals briskly and toughly with the headline parts of this thing, so that there is a much clearer separation between the headline stories and the background stories. So you look to David Brinkley, because, after all, he is news at NBC, and you pair him with somebody who talks very fast, somebody whose news presentation is extremely crisp. You separate it so David does the Washington stories and the somebody else does the other stories, the world and the nation stories, from New York. The two together are going to present a face to the world that says, "We're going to be efficient and fast. We're going to tell you in a weighted sense what's the most important, next most important and next most important story. We'll give some background, and then we're going to step back a little bit and take a look at something else. Then we're going to close the program. But its essence is going to be kind of a change of pace, in a very brisk sense." Think to yourself that it's the difference between a full-sized paper and a tabloid with a very carefully laid out front page, or a news magazine and a magazine of general interest. Then you might look for John Hart, or Jessica Savitch, or somebody who does that kind of thing in a very tough, straightforward way, because that's the personal characteristic that matches the kind of thing you want.

There was a time when we wanted to put on the Tom-Tom show, Tom Brokaw and Tom Snyder, and were seriously thinking of using them. I can now speak quite freely because I've left NBC and I really don't know what's going to happen or who will do what. But if that choice were yours, you'd have to begin to worry about the following things. You have a morning program which is relatively successful. Tom Brokaw is a central part of it. If you take Tom Brokaw out, who are you going to replace him with? You have a *Tomorrow* program, and a personality in Snyder. You've got to take him as he is, you can't change him to fit your needs, because over five nights a week, the way he comes across will be the way he really is.

Tom Snyder is a very electric person, but he has a sense, I think, that the news has weight and importance. He doesn't play around with the news. He'll never wear a funny hat on local news. He does odd things on the *Tomorrow* program, but he doesn't do them on news programs. He is a very good anchor person. His anchoring depends on a considerable amount of personal input, that is, he has to have some leeway in dealing with the thing. He doesn't deal with it by making jokes and wearing beanies and that sort of thing. He deals with it by talking about the news. Tom Snyder does something that's absolutely fascinating in television. He writes the way he speaks. You never hear Snyder reading from a teleprompter a phrase that sounds as if it was written. It always sounds conversational. That's a talent, and very few people can do that. Most people can talk fairly well if they have to talk, but if you ask them to write it out first (and you have to write it out for broadcast because you just can't skip it or get it wrong), it will sound written out. Cronkite sounds written out. He speaks differently from the way he appears on the air. Snyder doesn't. He's a very good anchor person.

Okay, then you have to decide if he is going to match the program. If he doesn't match the program, do you want to match the program to him? Would he pair off with Brinkley? And remember, Brinkley is really a very strong part of the entire operation. He is NBC to a big piece of your audience, and he's very good. You just don't want to lose him on plain esthetics. He has a terrific news sense. Brinkley's a guy who's watched the whole world go by and can give you very good advice on how to run a program. Well, are Snyder and Brinkley going to get along? Are Brokaw and Brinkley going to get along? How do they work out together as a balance? How does each of them match the kind of program you want to do?

The answer is that anybody who is competent to do the work can be an anchor, but not everybody can be successful. There have been unsuccessful anchors. Take Frank Reynolds, who is now doing ABC again. Frank Reynolds is competent to do the work, and he's as good as any other anchor person in any other place at any other time, given comparable circumstances. Your average, standard anchor person. The program wasn't successful with him in it, not because Frank Reynolds was bad, but because that intra-structure, that whole process wasn't as good as it should be.

The anchor person and the tone of the show have to be in balance. The funny thing is that people inside television don't understand that. They always think, "This guy's very jazzy." You always think that a local anchor person is terrific if he has a lot of hair, is relatively young, speaks clearly, and has good teeth. You think he's great, but you don't think he's so great on network. Why not? You begin to think on network that the news is more important, and on local it's not so important. On local, what are you dealing with but a bunch of sports? A guy who wears a funny-looking flower in his lapel is an asset in that sphere.

There's an interesting person named Ed Newman who doesn't have enough hair to cover a billiard ball. He really has a funny voice. He doesn't write conversationally, and yet he is a very good anchor. When he does an NBC news special, he's like a utility infielder. You have an anchored special, and everybody believes him. He violates all the possible canons, but he's very good. That's because, first, he is transparently intelligent. Second, he matches the matter he is dealing with. He's interesting and interested, and dealing with stuff in which you will be interested. In addition, he is comfortable with the material. He knows that he can handle it, that he can deal with it, and for those reasons he comes across well. He wouldn't come across well anchoring a local program. But, if you take a lot of local anchors and put them in the situation Ed Newman's in, I'd bet my own money they would be awful, because just having the hair, the teeth and the perfect profile doesn't make you an anchorman. It doesn't even make you an anchor on local news. But it tends to.

I am somewhat disdainful about some of the things local news people do because they are unreflective. Local news is fascinating because, to go back to what we were saying earlier, stations make a lot of money out of local programming. Stations can make a lot of money on local news. Therefore, there's a lot of time given to local news. Well, here you are, you're in the middle market—not the top ten markets, but somewhere between 10 and 50. Your station is going to make a lot of money on local news, and you've got about an hour to fill. Your budget isn't big enough to do what you really want, but it's pretty big. You've got enough people to help you.

What happens is they begin to invent ways to do things. Three quarters of television between 1948 and 1968 was invented by networks. Since 1968, three quarters has been invented by local news. This great electronic advance that we see in today's news coverage is really engineered best in local news, because they have the resources, the time on the air to play, the people who are willing to take chances, and they're not operating under such a spotlight. Since network news is such a question before the public, every change you make is national and is reported on. But if you're in Des Moines and you want to try somebody out, you put him or her on the air doing whatever, and if it doesn't work out, well there might be a piece in the paper about it, period. It's not such a big deal. So the experimentation, the new things involved, come in large number from local news. They are the people who pioneered mini-documentaries. They are the people who pioneered new ways of using electronic cameras, and they are active in these very vital, throbbing areas. The problem is that in their fights for the ratings, they tend to get bemused by what they *think* they should be doing, not what they really should be doing—they get these nutty ideas. A whole news division in a local station is doing very good work and, because "happy news" seems to be selling somewhere, they wind up with an anchor person who is notably short on brains, but he looks good and he makes an occasional written ad lib sound as though it wasn't all that written. It's a funny blind spot on the part of news directors, who are very bright people, and news staffs, who are very good. Too many times they're unwilling to take the chance on a guy who looks like Ed Newman, because God knows what will happen if the ratings go down.

QUESTION: How much writing do the anchor people like David Brinkley and Walter Cronkite do on the news?

RICHARD WALD: The ones I know best, of course, are the NBC people. By and large, major anchor people do their own writing. Brinkley, I think, rarely reads anything he hasn't written. Chancellor sometimes reads things he hasn't written, because it has to be written while he's doing something else. But other than those exceptions, they do all their own writing. The people on the *Today* program don't write. They do their program from notes on the teleprompter. It isn't exactly ad lib, but it isn't written out either. All the facts are laid out and they look at it and say what they feel would be appropriate. The same is true for Cronkite roughly. I don't know very much about how they handle the ABC program, but it's so complicated now that I think it needs a lot more supervision as to who writes what. I don't know if Harry and Barbara and Frank Reynolds write their own stories, but by and large, you get to a job like that by being competent enough to write it.

QUESTION: Isn't there a trend, at least on local news, towards people *not* writing their own stories? There was an article in the *Los Angeles Times* about stations going out and finding people who had nothing to do with news or journalism, who may not even have college degrees, no experience whatsoever with current events or news, and because of appearance or whatever, turning them into news people on local programs. I would assume they don't write their own material.

RICHARD WALD: It's very hard to write a television news spot, because it has to depend on what the film that accompanies it shows. I'm not talking about rewriting a piece of wire copy so you can read it on television. Something that goes with film is of some complication. It's hard to write because it has to be done in a fairly short time period, it has to be done after the cutting of the picture, and so forth. So that, for some people, especially on local news where you get 60 minutes of this stuff, minus commercial time, no anchor could write all of that. There isn't the time to do it. They need help, so there are writers. There is a tendency, in television, to pick on-camera people because they look good or sound good, sometimes both. It is nowhere written that because you look good and sound good you can't report news, or be an anchor. It's just that if your only qualifications are looking and sounding good, chances are you will not be terribly successful. Or, if you are successful, you won't be able to write it yourself. I am positive that there are successful people in radio and television who could not write a line of their own. But there are very few. There are none in network television news that I know of, and there aren't as many as you might think in local news, because the ones who can't do the work tend to fall of their own weight after a while. They aren't flexible enough to do anything but the very simple thing for which they were hired, and the news needs change all the time. So they find it very hard to bend with the changing needs and they go.

Outside Documentaries

SONNY FOX: It has always been criticized that the networks won't accept outside documentaries, that is, documentaries produced by outside producers. Would you explain why?

RICHARD WALD: It's a complicated process that hinges on the fact that you are legally liable for what you do on the air in a couple of ways. You can lose everybody's license if you do something terrible. Another reason is that you are subject to various rules like the fairness doctrine on equal time, that requires you to give time to both sides of an issue, and remember time is the essential thing in broadcasting. You can also lose your credibility, and credibility is the pillar on which news is built. If nobody believes it, they're not going to watch it. It has been the experience of the networks so far that news programs produced internally vary in quality but are usually accountable. You can answer to the fact that it was produced reasonably well under reasonable rules. And every once in a while there's an outside-produced package that brings with it difficulties that just reverberate forever. One quasi-documentary by an outside producer that NBC used purported to be in one part of the country and was in fact in another. It purported to show a particular kind of bear doing something, but in fact was showing something else. It really wasn't a terrible sin, but it was staging, and it's against the law. One news documentary that I remember, that was never shown on network television, was about a foreign country that was hard to get into, and was produced by a guy who was trying to negotiate financial arrangements with that country. I would fire anybody on a news staff who was producing a program and at the same time trying to arrange contracts with the country. You can't do that, whether or not you're being honest. The appearance of impropriety is too great.

SONNY FOX: And yet, the *Today Show,* produced by the NBC news operation, continues to accept transportation and lodgings when it travels abroad, or to other parts of the country, and originates from those areas.

RICHARD WALD: It does so now under certain clearly defined rubrics. It does no longer accept, because I really didn't like it and managed to stop it, various gratuities that it used to accept. It did in the past, to the great discomfort of everyone and with as many disclaimers as possible. There was always the general sense that we didn't like it and we would not be bound by it. Everybody inside knew it, as well as everybody outside. There was no fiddling around about it. We were taking freebees, and it really meant that when you got to those foreign countries you would bend over backward to do something that would embarrass them. That's why we never went back again.

The policy of accepting freebees irritated a lot of people. It irritated me too. I don't like it, I didn't like it, and I stopped it, but it's one of those things that's very hard to stop, because once it gets started it goes on. Another thing the *Today* program was guilty of, a very difficult thing to persuade television people we shouldn't do, was having the news people read commercials. That really was unpleasant to those of us who didn't like it, and we managed to stop it.

Production Costs

QUESTION: The cost of production for a two-hour television movie runs about $1.1 or $1.2 million. By comparison, how much would a one-hour prime-time news special cost?

RICHARD WALD: They range, but the average cost is somewhere from $250,000 to about $300,000 for each of the three networks. Part of that range is involved in the accounting procedures. If you use network equipment and everything else, the charges are very high because they charge off man-hours against the programs using it, following all sorts of complicated systems. But I think in real dollars it would come out to approximately half to two thirds the cost of an entertainment hour, and documentaries are seldom repeated. They don't recover much of their production cost in syndication or repeat showings.

Slanting the News

SONNY FOX: How do you answer critics of the network news who charge the news with being slanted, or who challenge the power that resides in a few people to decide what the American people will, by and large, watch on the news?

RICHARD WALD: Insofar as slanting is concerned, I don't know of any. I've never seen any. The best argument you can make against us is that we do it subconsciously, that we can't even recognize it when it's pointed out to us. If you can tell me that, there's no way I can answer. If I am so dumb or slanted that I don't understand it when somebody says it to me, or see it when it's pointed out, I guess maybe I do it. But if I have any conscious competence, I don't.

As to the question of the power of television news, I think there's a funny problem in that. Nobody knows what is the power of television news. There is no study I've ever heard of. Everybody intuits that it's powerful. Everybody says, well, you see it all over and you get to everybody at once, and therefore it's powerful. Nobody knows. If we were all busily slanting the news, and we were so powerful, and our critics swear all three networks were against Nixon, then why didn't we prevent his getting elected in 1972? Television's powerful, and they say we were all slanted against Nixon, so he couldn't have gotten elected! It just isn't true. The observable, pragmatic truth is that television is an interesting, experiential thing that tends to produce a reaction in people, but the reaction is not predictable. It is so complex, just as human beings are complex, that it is very hard to know what it will do *to* you or *for* you next.

A very subtle problem with television is that the critics construct great walls to keep it in, because it is such a powerful beast, and they don't know how it will work. The people inside television tend to internalize that idea so they become afraid of this imputed power, and what they begin to do is to create a cage for themselves. People will tell you, and you can observe for yourself, that the amount of controversial material available on any given television documentary is minute. It's bland. Not terribly exciting. Why is that? The answer is that an artificial set of foolish doctrines promulgated by well-meaning people creates a funny, mathematically even-handed approach to things. But, even more, there's a sense among people in broadcasting generally, and television in particular, that they have to be very careful lest they loose the demons of opinion and prejudice upon an unsuspecting world. This feeling creates a calcification, a kind of internal rigidity that does not allow them to appear to be interested in or worried about the subjects on which they report. And the result of that is that every night you can see television news that seems to be impersonal. The reason you forget the stories on television as quickly as you do, and you do, is that the stories themselves have no weight or particular virtue. They are abstracts constructed by people who are worried about fairness and balance, and also worried lest they show an interest in the subject itself. And the more abstract, the easier it is to forget. The truth of our lives is that those subjects in which you become interested are terribly complicated, and terribly human. On the one hand television is really human and experiential in what it does, but television news tries to be holier than anybody else and purer than pure, more balanced than balanced, and the two things don't mesh well. There's a real contradiction in the way television news is done. I do not believe that television news is biased. It isn't. But the way you see it may be biased.

The best possible example I can think of came at the time of the Kent State shooting. There was a girl killed whose father lived in Pittsburgh, and he had a morning press conference the next day on the front lawn of his house. He stood on his lawn and made a statement to a massed phalanx of television cameras. My network, like the other two, was inundated with outraged people who said, "How dare you judge what happened in Kent State when you don't really know! Why did you let that grief-crazed father say those terrible things against the President? Why did you let him condemn the National Guard?" And it was almost universal among people who were in favor of the administration and against kids with long hair that this man had done something awful and only the prejudice of news divisions allowed him to do that. I myself was so shattered by this, our blatant prejudice, that I went over the record. What he said, essentially, was "Mr. President, I am a Republican, and I voted for you. My daughter is now dead in Kent State in Ohio. Could you appoint a commission that would find out why? I believe in this country, I believe in the Republican party. Something has gone wrong. Can you tell us what it is?" A very moderate, a very pro-Nixon statement, but because of the time, because of the way he was crying, because of the effect he had upon the audience, those people who were worried about the effect of Kent State assumed he was saying things that would be anti-administration. It just wasn't true. But in your stomach, you don't think. In your stomach you just receive. And television, a lot of the time, speaks to your stomach. There's no other way to explain it.

Ratings

QUESTION: How much do the ratings influence the educational or sensational nature of the national news?

RICHARD WALD: I can't give you a precise answer, but you must keep in mind that television is a mass medium, and the size of the mass is of importance to the success of the project itself. It needn't be enormous, but if it is absolutely minuscule, you have wasted the resources of the medium itself. You might just as well be handing out leaflets. So that the size of a television audience is necessarily larger than the size of other possible media audiences. That said, there is an enormous amount of leeway in what you can and will and should do. Networks, oddly enough, are not looking for large audiences in most of the special news programs they do. They are not looking for it because they assume immediately that, whether the rating is high or low, it's going to be lower than entertainment. They'll lose money on it anyway, and they really don't give a damn whether it's a 2 or a 12 or a 24 rating. They'd like it to be 24. It's always nicer if it's higher, but essentially, once they've decided to make that sacrifice, they're willing to accept that sacrifice.

However, in a regularly scheduled program, which is competing with other regularly scheduled programs, they want to get a larger audience. Why? It is in the nature of most news operations, and we're talking about mass media now (whether it's magazines, newspapers, radio stations or television), to be competitive. If you are doing as good a job as you can, more people will watch you. It is in the nature of both the commercial enterprise, which wants to make a lot of money out of it, and the creative enterprise, which will get more satisfaction out of it if more people watch you than watch the other fellow. So there is an enormous amount of attention paid to those things that will increase the audience. If you are a news purist, what you want to do is tell the truth. If you want to tell the truth, you want to tell it to as many people as possible. If you're just in it for the money, you want to get to as many people as possible, because the more people, the more money. So both interests coincide.

The big question that is always asked is, "Does that coincidence of interest lead to distortion?" And the big criticism made against televison news is that it does. I don't think it does. People think that things are distorted because you didn't put on as much about the gold crisis as you did about a hippopotamus lost in Los Angeles. Well, you can't do it. If you put on the gold crisis information, not only will people not watch, but those who watch won't get as much out of it as they would out of something printed. Whereas, the hippo will interest a lot of people. It isn't world-shaking, but it is of interest.

There's a great rule in all news operations, which is often misunderstood, and it goes: "Interesting is also important." If the entire country is mesmerized by the Beatles, then the Beatles are important. You cannot run a news operation and then say, "We're not going to pay any attention to the Beatles, because they are not important." They are important. They may not be of the order of importance you would like, they may not be on the same level of importance as the recognition of Red China, but they are important. They're important to the way we live. I do not believe there is a distortion of television news in order to get ratings. But clearly, there is a shaping of television in order to get ratings, and clearly also, the people involved in the news production process are interested in the ratings. It is a question of balance. The question always is, will you put something on that may not get ratings, but may be important? And so far, television news, like other news organizations, has always said yes. I know of no network programming or local programming which would ignore a story simply because, although they thought it was important, they thought it might cost them rating points. They may not make it the lead, but they'll put it on, and I think that's the kind of test you have to apply.

Index

Recordings of the lectures which form the basis for the chapters in this book are available in audio cassette form from Jeffrey Norton Publishers, Inc., 145 E. 49th Street, New York, N.Y. 10017. The cassettes are sold as a full 8-part series, or as individual lectures. For further information write to Jeffrey Norton Publishers at the address given above.